THE SEARCH FOR

ALIENS

ROUGH GUIDES

www.roughguides.com

Credits

The Search for Aliens

Editing: Tracy Hopkins
& Richie Unterberger
Layout: Sachin Gupta
Diagrams: Katie Lloyd-Jones
Proofreading: Jason Freeman
Production: Rebecca Short

Rough Guides Reference

Editors: Kate Berens, Tom Cabot,
Tracy Hopkins, Matthew Milton,
Joe Staines
Director: Andrew Lockett

Publishing information

This first edition published January 2012 by
Rough Guides Ltd, 80 Strand, London WC2R 0RL
11 Community Centre, Panchsheel Park, New Delhi 110017, India
Email: mail@roughguides.com

Distributed by the Penguin Group:
Penguin Books Ltd, 80 Strand, London WC2R 0RL
Penguin Group (USA), 375 Hudson Street, NY 10014, USA
Penguin Group (Australia), 250 Camberwell Road, Camberwell, Victoria 3124, Australia
Penguin Group (New Zealand), 67 Apollo Drive, Mairangi Bay, Auckland 1310,
New Zealand

Rough Guides is represented in Canada by Tourmaline Editions Inc.,
662 King Street West, Suite 304, Toronto

Printed in Singapore by Toppan Security Printing Pte. Ltd.

272 pages; includes index

A catalogue record for this book is available from the British Library

ISBN: 978-1-40538-324-0

1 3 5 7 9 8 6 4 2

THE SEARCH FOR
ALIENS

A ROUGH GUIDE to LIFE
ON OTHER WORLDS

by
Piers Bizony

www.roughguides.com

For Roger Scotford, who doesn't believe in them.

Contents

Acknowledgements

I would like to express my gratitude to a number of kind and clever people who assisted me with this project, and took the time to answer my incessant queries. Dr Everett K. Gibson and Dr David S. McKay at NASA's Johnson Space Center in Houston told me about Martian meteorite ALH84001, while Monica Grady, Professor of Planetary and Space Science at the Open University, patiently taught me the difference between one lump of meteoritic rock and another. Roger Launius, chief space historian at the Smithsonian Institution in Washington, DC, helped me check the history of project Viking, and Dr Gilbert Levin of Biospherics Inc., told me, with undiminished enthusiasm, about his life-detecting experiment on that spacecraft. Doug Millard, curator of the space gallery at the Science Museum in London, was an ever-reliable source of guidance, and Dr David G. Stork, at the Ricoh Research Center, Menlo Park, California, gave me some great ideas about swarms of robotic probes sweeping across the galaxy…

Peter Tallack at the Science Factory literary agency helped set this project in motion, knowing very well that the subject of extraterrestrial life is close to my heart. Andrew Lockett at Rough Guides was a cool-headed editor, guiding me painlessly towards what was needed. Many thanks, also, to Tracy Hopkins, who worked with me on the manuscript and spotted murky patches in my reasoning before they got out of hand. As ever, my family, Fiona, Alma and Oskar, were patient with me as I muttered about the radial velocity method or the hydrogen line while there was washing-up yet to be done. My excuse, "Why bother about dirty plates when we have the stars to explore?" didn't work. So thanks, finally, to Sugar, who licked them spotlessly clean for me. When you want a job done properly, get a dog.

Introduction

There are some questions we just can't help asking ourselves. Is our path from birth to death predetermined by inexorable laws of physics and biology, or do we have some control over our own destinies? Is there anything special about consciousness, or is it just nerve cells and knee-jerk spasms? And are we alone in the universe, or is there someone else out there who shares our perplexity and wonder?

On and on, the questions come. On a good day, the universe seems imbued with a kind of magic. The stars glow in the night sky like beacons of optimism. Surely this great, beautiful canopy of Creation is teeming with other life-bearing planets? Is it so unreasonable to suppose that, among that multitude of alien worlds, at least one might have given rise to someone – or something – intelligent? On a bad day, all of existence seems like motiveless motes of dust, with our small, insignificant ball of rock, the Earth, drifting in an immense void. Our instincts rebel against such loneliness.

Even the supposedly dispassionate scientists can't settle such deep problems. Some say that atoms and molecules, and calculable physical forces, account for everything, including art, poetry and love. Others insist that it's not the individual components of the material world that count so much as the intangible ways they work in harness with each other. There's a universe made of subatomic components and pieces of mass and energy, or else there's just one entity and all its energies and events are inextricably bound up with all the others, no matter how widely distributed. Who knows?

We congratulate ourselves for unravelling so many of nature's individual mysteries, from DNA to the birth and death of stars. At the same time, we find ourselves swamped by incomprehension about the broader framework. What we'd love to do is ask someone else – someone who lives in another part of the universe, and perhaps sees things differently (or even sees different things). Our desire for interstellar conversation is phrased in the technical language of radio astronomy, molecular biology, cosmology and other specialisms of our technological age, yet it all boils down to that one recurring question: what does it mean to be alive? This Rough Guide can't answer that question. Instead, it looks at how we might get that conversation going in the first place.

Piers Bizony

How this books works

This book comprises fourteen chapters. The first, **A plurality of worlds**, looks at the history of speculating about the existence of other worlds, dating back as far as our historical understanding allows, to the time of the ancient Greeks. Next comes **Aliens of the imagination**, which takes us to the dawn of the modern technological era, and the influence of early science fiction on the nascent quest for life elsewhere in space. **The origins of life** brings us back to the one world where life indisputably exists: the Earth. Before we can look for alien life, it's wise to understand what that word means in the first place.

Chapter 4, **The hunt for life on Mars**, shows how our basic assumptions about terrestrial biology were transplanted into the robotic assemblies

The sun allows NASA's Mars exploration rover Opportunity to take this self-portrait using its front hazard-avoidance camera.

of the first Mars landers in the 1970s, with teasingly ambiguous results. In **Mars reawakens,** a Mars rock lands on Earth, bringing some of the most intriguing clues ever encountered about Mars's early history, and its potential suitability for life long ago. Moving on from the Red Planet, **If not Mars, where?** looks at other planets and moons in the solar system that might be better candidates for life. It also explores the role of asteroids and comets in the pre-biological development of planets.

The cosmic haystack, Chapter 7, is where we turn our attention far beyond our own solar system, and look at the surrounding galaxy, the Milky Way. All those distant stars and alien solar systems are too far away for us to study, except by radio and other forms of electromagnetic radiation, or "light" as it's more commonly known. So, what clues should we look for in our search for life in the galaxy? **Listening for a signal** delves into the specifics of SETI, the Search for Extraterrestrial Intelligence, and introduces many of the people who do the listening. And in **The planet hunters,** it's not intelligent messages so much as potentially habitable planets that the experts seek, via the latest developments in optical astronomy.

Chapter 10, **Intelligent civilizations,** asks: What might be the nature of an alien society? Would it share any of our intellectual and emotional aspirations? **The long road to the stars** looks at the daunting problems of interstellar travel, while **Kidnapped by aliens** tries to understand where all those flying saucer abduction stories come from. Taking a step back, **The bigger picture** considers the entire fabric of existence, and our place within it. Are we just specks of dust in an uncaring infinity, or is there something special about the fact that we're aware of that great vastness? Finally, the **Resources** chapter directs you to useful books, websites, documents and one or two very special alien-related movies.

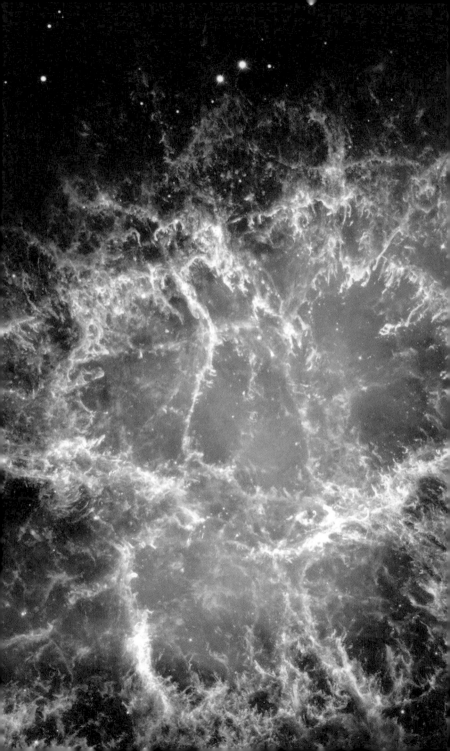

A plurality
of worlds

At first glance, our fascination with the possibility of life on other worlds seems like a contemporary strand of thought, bound up with the recent arrival of rockets and space technology. In fact there's never been a time when humans were content just to settle for this one world.

Our ancestors had their spirit realms, non-corporeal zones of life co-existing alongside the immediately familiar world of animals and plants, mountains and streams, yet inaccessible to human senses except through shamanistic rituals. Even today, most religions believe in a division between "this" world and the "next", where invisible gods are in charge. Many of us still hope that a higher realm of existence, some kind of heaven or paradise, awaits the souls of the honourable dead, the saints and the warriors, while the less honourable deceased end up in, shall we say, a lower plane.

The sky has always beckoned, with its seemingly magical cycles of repeated patterns and movements. Ancient peoples saw the sun rising and falling with inexplicable regularity, while, measured against landmarks on the horizon, the gradual drift and return of constellations foretold the seasons for those who knew how to decrypt their intricate messages. Four thousand years ago, the Mesopotamian civilizations, in what is now Iraq, could navigate by the stars and predict the movements of five planets. In the absence of telescopes and scientific theories, some of those long-vanished peoples must have wondered about the bright dots of light in the night sky, and whether or not any of them might be inhabited. There were gods and spirits aplenty up there,

> "It is extremely unlikely that this earth and sky is the only one to have been created. In other regions, there are other earths and various tribes of men and breeds of beasts."
>
> Lucretius (99 BC–AD 55), Roman poet and philosopher

but could anyone living on Earth during the first few millennia of human civilization grasp the physicality of what we now call planets and moons, suns and solar systems?

Among the ancient Greek thinkers, we are confident that Thales of Miletus (624–546 BC) and his student Anaximander (610–546 BC) were among the first philosophers to talk, in a non-spiritual way, of tangible, inhabitable worlds other than our own. Around 500 BC, a new idea made the rounds among the most radical Greek philosophers: *aperoi kosmoi*, the plurality of worlds. Today we call this **pluralism**.

The roots of pluralism

The Greeks didn't base their ideas on anything so vulgar as scientific observation. Their well-bred minds were for thinking, while lowly servants, slaves and women did the doing. Even so, clear-headed logic got them quite a long way. Metrodorus (fourth century BC) asserted: "Things that exist tend to do so in great multiplicity, from grains of sand to ears of corn. It would be odd if only a single stem of corn grew in a field. Similarly, it would be strange if our Earth were the only one in the cosmos." Metrodorus's argument seems just as compelling today as it did 2600 years ago. Later on, we'll encounter its modern formulation, the Mediocrity Principle (see p.29).

Epicurus (341–270 BC) thought the purpose of philosophy was to help us achieve happiness and tranquillity. (Modern Epicureans love their pleasures, especially a fine meal.) So generous was Epicurus's worldview that he thought it must apply "to infinite worlds, both like and unlike this world of ours". The great philosopher Aristotle (384–322 BC) didn't want to share the goodies quite so far and wide, so he put the brakes on pluralism. The Earth, he insisted, must be at the centre of the cosmos, itself consisting of fifty or so crystalline spheres, nested one inside the next like transparent layers of onion skin. The sun, moon, planets and stars were studded on the surfaces of those spheres like glow-worms trapped in glue.

Annoyingly, sometimes the planets appeared to travel backwards across the night sky, like rebellious glow-worms breaking free of their sticky bonds.

> **"How unreasonable it would be to suppose that, besides the earth and the sky which we can see, there are no other skies and no other earths."**
>
> Teng Mu, Sung Dynasty sage, thirteenth-century China

The interlaced cogwheels of the Ptolemaic system put the Earth at the centre of the cosmos, with the sun and all the planets orbiting around it.

Today we know that these cosmic illusions are generated by the relative motions of the Earth and planets along their different orbits. Around 1870 years ago, a quick and dirty fix was needed to clean up the calculations. The Greek mathematician Ptolemy (AD 90–168), based in the Egyptian city of Alexandria, reattached the wayward wanderers using **epicycles**, smaller crystal spheres on the periphery of the main ones. A cross-sectional slice through this arcane geometry looks rather like a maze of interlaced cogwheels in an overambitious clock. Ptolemy's complicated model dominated astronomy for the next 1400 years. Between them, Aristotle and Ptolemy created a universe in which the Earth – and by extension, humanity – was the lynchpin of existence.

As the great Graeco–Roman civilizations fell apart and Europe entered the Dark Ages, Islamic scholars and merchants from the Arabic world reverently preserved the intellectual legacy of ancient Greece. When these merchants came to Europe, carrying those ideas with them, they inadvertently kick-started the Renaissance, Europe's eager (if slow and painful) rediscovery of its mislaid classical inheritance. However, as Christian dogma took hold around the same time, the Catholic Church of Rome established itself not merely as an engine of spiritual beliefs but a font of earthly authority for kings and cardinals alike. It became not just absurd but dangerous to suggest that the human realm could be anything other than the very heart of Creation. Aristotle and Ptolemy were absorbed into the Vatican's frames of reference – not because they were intellectually unassailable, but because they helped justify the Church's centrality in human affairs.

Questioning the Earth-centred universe

During the sixteenth century, European astronomers continued to have great respect for Ptolemy as a pioneer of their science, even as problems began to emerge with his supposedly brilliant scheme. Measurements of the night sky became more accurate, and countless new observations refused to fit into the established theory. Errors accumulated in the Ptolemaic system like damp patches on a ceiling. A better theory of the heavens was needed.

In 1543, the last year of his life, Polish cleric and astronomer **Nicolaus Copernicus** (1473–1543) thought it was as good a time as any to publish his dangerous new book, *De revolutionibus orbium coelestium* ("On the Revolutions of the Celestial Spheres"). He reasoned that all of the flaws in Ptolemy's concept could be explained by assuming that the Earth and planets orbited the sun. He was within a hair's breadth of a theory that worked, except for his insistence that, given God's perfection, all celestial movements had to be perfectly circular. German mathematician Johannes Kepler (1571–1630) added the last crucial refinements to the scheme by number-crunching the data from Tycho Brahe (1546–1601), a Danish astronomer who made extremely accurate observations of planetary movements. Kepler concluded that planets orbit in egg-shaped ellipses, rather than circles. This turned out to be bang on the money. By the 1620s, the underpinnings of modern astronomy were taking shape – even if the notion of God's supposedly perfect celestial neatness had to be discarded. The watchful Church authorities insisted that these new calculations should be regarded as useful geometric abstractions, and nothing more. Heaven help anyone who claimed they mirrored reality.

Galileo Galilei (1564–1642), the founder of modern experimental physics, did indeed make such a claim, and heaven most assuredly failed to come to his rescue – although he was, at least, spared the worst fires of hell. In 1610, Galileo began experimenting with a new instrument, today commonly known as a telescope. Although he didn't invent this device, he came up with a decent model as soon as he heard that craftsmen in the Netherlands had worked out the basics. We can be reasonably sure that he was the first person to point this new instrument into the night sky and draw startling conclusions based on what he found up there.

Galileo saw Jupiter's moons, which was a stunning feat in itself. Of even greater significance was his observation that sunlight sometimes illuminates Venus full-on and at other times obliquely, so it appears to gradually change shape from a full circle to a thin crescent and back again, as is the case with our own moon. It was obvious to Galileo, therefore, that Venus was a sphere, a tangible place, and not merely an abstract dot of light in the sky. It was also obvious to him that the sun couldn't be orbiting the Earth. Rather, the Earth, and all other known planets, had to be orbiting the sun. This could only mean that the Earth was not the centre of the universe, an idea quite literally unthinkable to the Church. In 1633, Galileo was threatened with torture and execution for heresy, and was persuaded to withdraw his claims — in public, at least. He spent the rest of his life under house arrest, although his great renown across Europe helped save him from contact with the torturer's tools.

Galileo demonstrates the telescope to the Senate of Venice, and later discovers that the Earth is not the centre of the universe.

Inhabited worlds

On learning of Galileo's discoveries, Kepler reckoned that the moons of Jupiter must be visible in the night sky specifically for the benefit of Jovian inhabitants. Foreshadowing a very modern strand of speculation about advanced alien civilizations, he even suggested that other intelligent life forms might gaze on the Earth from afar and find it "more amenable to their understanding than to ours". Galileo had little time for talk of life on other worlds, let alone of beings superior to mankind. In 1613, he wrote of "the false and damnable view of those who would put inhabitants on Jupiter, Venus, Saturn and the moon ... animals like ours, and men in particular". He and Kepler corresponded in quite a friendly manner, although Galileo was scornful towards any rival's ideas that didn't accord with his own. Perhaps his contempt in this case was also directed towards Italian astronomer Giordano Bruno (1548–1600), arguably the first man to be executed for weaving aliens into his heresies.

Bruno began his career as a scholar at a prominent Dominican monastery in Naples, but swiftly drifted off-message, especially when abroad on lecture tours. In 1591, the notorious Catholic hit squad known as the Inquisition caught up with Bruno in Venice and hauled him off to Rome, where he was kept imprisoned for eight years, until the day came when the authorities decided to silence him for good by burning him to a crisp. Historians are not sure exactly why he was so cruelly punished, but among his many heresies, his startlingly prescient ideas about extraterrestrial life must have made uncomfortable reading for the Church. In *De l'Infinito, universo e mondi* ("On the Infinite Universe and Worlds"), published in 1584, Bruno defended the sun-centred theories of Copernicus, and described "countless suns and countless earths all rotating around their suns in exactly the same way as the seven planets of our system". He thought that other worlds in the universe must be "no less inhabited than our Earth", and urged the destruction of "theories that the Earth is the centre of the Universe!"

Bruno's fellow countryman, philosopher and astrologer Tommasso Campanella (1568–1639), played both ends against the middle by maintaining that nothing in the Bible specifically forbade the existence of other inhabited worlds. As the Protestant cause took hold

> "On some worlds there is no Sun and Moon. Others are larger than our world. There are some worlds devoid of living creatures or plants or even moisture."
>
> Democritus (460–370 BC), Greek philosopher

in the seventeenth century, pluralism became a useful weapon in the anti-Catholic armoury, even as the theory of a sun-centred universe became commonly accepted. In 1657, Pierre Borel (1620–71) – Protestant thinker, keen chemical experimenter and physician to King Louis XIV of France – published *Discours nouveau prouvant la pluralité des mondes* ("A New Treatise Proving a Multiplicity of Worlds"), in which he suggested that the best way to discover life on the moon or on other planets would be to visit those worlds via "aerial navigation".

Borel was inspired by René Descartes (1596–1650), one of the first great European intellectuals to describe the physical universe as a mathematically self-consistent engine of matter and motion. Descartes suggested that great swirling motions, or vortices, are responsible for the orbits of the planets around the sun, or of moons around planets. Heavier worlds travel near the outside of their particular vortex, while lighter bodies remain closer to the centre. It was a flawed scheme, but Descartes presaged at least some of Newton's ideas about gravity by several decades.

More to our point, Descartes assumed that all stars are like our sun, and that each and every one of them must have its own retinue of planets – and he could see no reason why those worlds shouldn't be inhabited. Today we understand something that he couldn't possibly have known. Stars come in different varieties, from young, super-hot short-lived monsters crowded together in stellar nurseries to small, dim, brown dwarfs slowly ageing in isolated obscurity. Longer-living stable stars like our own sun occupy the middle ground. Only a certain proportion of stars in the universe might be suitable hosts for life-bearing planets.

A more enlightened otherworldly view

Even so, the cosmological ideas expressed by Descartes and his followers have many parallels to modern theories. In 1745, another of those busy French intellectuals, naturalist George-Louis Leclerc (1707–88), suggested that ancient comets smashing into the sun had dislodged chunks of its surface, which then condensed to become the planets. It wasn't quite the right idea, but was creditably close to our current understanding of protostars and planetary accretion discs.

Throughout the eighteenth and nineteenth centuries, the profession of astronomer became firmly entrenched, largely funded by the urgent navigational requirements of merchant explorers and imperial navies. Even though Isaac Newton (1643–1727) had a secret soft spot for alchemy, the

Aliens and the problem of the soul

The Church of England has always been amenable to the intellectual hobbies of its servants. Clever reverends haunting the corridors of Oxford and Cambridge were allowed leeway for wacky cosmological ideas. John Wilkins (1614–72), bishop of Chester and respected scientific thinker, was just one of many churchmen to chatter about life on other worlds, both in print and at table, without fearing the flames of damnation. By now, the main cause for complaint wasn't that other worlds in the universe were impossible, so much as the questionable state of grace that the inhabitants of such worlds – if they existed – might enjoy. Could they sin and be redeemed, or was this a privilege set aside for humanity alone?

It's easy to think that this sort of problem would be laughable if anyone worried about it today. Far from it. The mainstream Christian establishment accepts that twenty-first-century science heralds the possibility of other intelligent and self-aware creatures existing elsewhere in the universe. In 2008, the Pope's astronomer, José Gabriel Funes, said that "Just as there is a multiplicity of creatures on Earth, so there could be other beings created by God". He was vague, however, on the precise details. For instance, must each world have its equivalent of Jesus Christ? If so, then how can each one of these alien redeemers be *the* son of God? And does each sentient race have its own heaven? The discovery of extraterrestrial signals or other traces of an alien civilization would throw the Christian Church, and many other established religions, into theological confusion.

rigour of his mathematical physics swept ancient mysticisms into disuse, along with the Ptolemaic system and the Church's powers to complain about anything to do with the sky. Some of Europe's best minds began to realize that the inhabitants of other worlds were unlikely to share such human concerns. According to Gottfried Leibniz (1646–1716), Newton's rival in the creation of calculus, "It must be acknowledged that there is an infinite number of globes that have as much right as ours to hold rational inhabitants, though it follows not at all that they are human."

The Jewish faith has never felt as threatened by the possibility of alien minds. The influential Talmudic scholar Rabbi Judah ben Barzilai of Barcelona (early twelfth century AD) made the delightfully specific assertion that 18,000 other worlds should support intelligent beings. Islamic scholars are also amenable to the idea of God distributing the seeds of sentient life more widely than on our world alone. A verse in the Koran tells us, "Among His signs is the creation of the heavens and the earth, and the living creatures that He has scattered through them." The complexities of Hindu scriptures allow for a plurality of worlds, although it can be hard to distinguish mystical planes of existence from physical ones. The world's faiths, myths and traditions are rich in pluralistic allusions, but the belief in extraterrestrials as tangible, physical entities has its roots in European rationalist thought.

As the Age of Enlightenment segued into the Industrial Revolution, the notion of life on other worlds was transformed from mere fanciful philosophizing into something we could actually expect to look for, provided that the right instruments or experiments could be devised for the hunt. The only remaining hurdle was to phrase the quest in scientifically credible terminology, so that funding applications could be taken seriously. As is so often the case in technological history, only a few space boffins at the dawn of the twentieth century had a knack for selling their cherished ideas to a wider audience. Instead, most of the marketing was done by scribblers of trash and nonsense, pedlars of pish and piffle, and even one or two merchants of twaddle: the hack writers and pulp publishers who devised the golden age of science fiction.

Inventing a signal

Attempts to signal our presence to whomever – or whatever – might be out there date back at least to the eighteenth century. The German mathematician Carl Friedrich Gauss (1777–1855) thought that huge steerable mirrors could be used to reflect sunlight towards the planets, in order to send simple pulses proving the intelligent occupation of our planet. Alternatively, he proposed chopping a giant Pythagorean (right-angled) triangle of clear land out of the Siberian forest and sowing it with bright-coloured wheat so that any intelligent aliens who pointed their telescopes towards us could admire our grasp of geometry. Austrian astronomer Joseph von Littrow (1781–1840) had a scheme to pour kerosene into a 30km-wide circular canal and light it at night to signal our presence.

Two famous technology pioneers showed early interest in interplanetary communication via radio. In 1901, Serbian-born electrical engineer Nikola Tesla (1856–1943) reported receiving a strange signal, possibly from Mars, on his giant transmitting tower in Colorado Springs. Nineteen years later, Italian physicist Guglielmo Marconi (1874–1937) told reporters he had detected "very queer sounds and indications, which might come from somewhere outside the Earth". Tesla and Marconi's less excitable scientific contemporaries dismissed all these signals as spurious interference from nearby telegraph systems, and the story faded.

For most scientists in the early twentieth century, optical light, rather than radio waves, still seemed the most practical way to reach out to aliens. After World War I, the US Sperry Gyroscope company wondered if hundreds of spare anti-aircraft spotlights might be gathered into a giant array to send pulses of light towards Mars.

Aliens of the imagination

Martian princesses, sentient creatures in the oceans of Venus, eerie alien machines... twentieth-century science fiction shaped our expectations of space at a time when the solar system was still largely mysterious to us. Mars in particular influenced the imaginations of fiction writers and scientists alike.

The Martian landscape

When Italian astronomer Giovanni Schiaparelli (1835–1910) drew the first map of Mars in 1877, he had nothing more to guide him than the blurred images captured by his simple telescope. He recorded, as best he could, several large dark plains loosely connected by much narrower features, which he labelled *canali*. In Italian, this simply means grooves or channels. An enduring scientific myth was created when Schiaparelli's

Schiaparelli's map of Mars showing the channels (*canali*) that were mistaken for water-bearing canals created by an alien intelligence.

work was translated for English-reading astronomers. **Canali** was interpreted to mean "canals" – artificial water-bearing structures created by an intelligent civilization.

In 1894, misguided American astronomer Percival Lowell (1855–1916) began intensive observations of Mars from a high-altitude telescope observatory in Flagstaff, Arizona, which he built largely at his own expense. He was sure he could see Schiaparelli's *canali* criss-crossing the entire Martian surface. In 1896, he published his findings in a book (the first of three on a similar theme) simply entitled *Mars*: "On Mars we see the products of an intelligence. There is a network of irrigation … Certainly we see hints of beings in advance of us."

We'll never know what Lowell might have seen if he'd never encountered those flawed translations of Schiaparelli's work, and added some confusions of his own. We do know that Lowell's work was not seriously regarded by other astronomers. James Keeler, an astronomer at the Lick Observatory in California, politely suggested at the time that "the result of [Lowell's] ingenuity is that fact and fantasy have become inextricably tangled". Other critics were not so polite. In 1905, an article in the *British Journal of Medical Psychology* suggested that Lowell might have been influenced by "voyeuristic impulses" and "unresolved Oedipal complexes".

A new world: the dawn of science fiction

Even so, Lowell's erroneous concept of an ancient water-starved civilization girding an entire world with canals and pumping stations was much too entertaining to waste. Newspapers and popular writers took up Lowell's ideas with gusto – not least **H.G. Wells** (1866–1946) in his classic novel *The War of the Worlds* (1896), the famous tale of an invasion of Earth undertaken by intelligent but merciless Martians. The novel's opening pages retain all their power today:

> "No one would have believed in the last years of the nineteenth century that this world was being watched keenly and closely by intelligences greater than man's and yet as mortal as his own … Across the gulf of space, minds that are to our minds as ours are to the beasts that perish, intellects vast and cool and unsympathetic, regarded this Earth with envious eyes, and slowly and surely drew their plans against us."

By the dawn of the twentieth century, astronomers attached refracting prisms to the ends of their telescopes so they could read the various spectra of light emitted by stars or bouncing off the surfaces of planets. Analysis of sunlight reflected from the Martian surface, or glancing obliquely through its atmosphere, uncovered some sobering data. The air was whisper-thin and incredibly cold, and consisted mainly of carbon dioxide. It was all but impossible for liquid water to exist on the surface, whether in naturally formed streams or those great alien canals. Water could only exist as sheets of ice or permafrost. Given that liquid water has always been regarded as essential for life, these icy conditions seemed to present too harsh an environment for anything above the simplest bugs or lichens. The Martian master builders retreated from the scientific horizon, and made their last redoubt in the pages of fiction.

Aliens in the golden age of science fiction

Wells is regarded as the father of modern science fiction, but his most famous work was intended as an allegory of human folly and warfare rather than a literal exploration of alien life and its possible intentions towards us. Another British science-fiction pioneer, Olaf Stapledon (1886–1950), was more interested in genuine speculation about life on other worlds. His 1930 novel, *First and Last Men*, was widely respected

Martian princesses

The popular early twentieth-century image of aliens was essentially interchangeable with Martians or their close cousins, Venusians. Edgar Rice Burroughs (1875–1950), creator of the jungle hero Lord Greystoke (otherwise known as Tarzan), wrote a series of fantasies set on Mars, or "Barsoom" as its inhabitants preferred to call it. Like Wells before him, Burroughs was influenced by Lowell's ideas. Through his cheap and cheerful stories, a wide readership soon developed a taste for life on the Red Planet.

Barsoom's canals were the least of its attractions. The planet also boasted Dejah Thoris, princess of Helium. On meeting her, Burroughs' human hero, John Carter, could scarcely restrain his hormones. "The sight which met my eyes was that of a slender, girlish figure, similar in every detail to the earthly women of my past life", he noted. Her skin was of a light reddish copper colour, against which "the crimson glow of her cheeks and the ruby of her beautiful lips shone with a strangely enhancing effect". Significantly, she was "as destitute of clothes as the green Martians who accompanied her".

"A lank tentacular appendage gripped the edge of the cylinder..." *The War of the Worlds* (1906 edition).

by philosophers and literary figures at the time. This "future history" spans billions of years of human evolution via genetic engineering and artificial intelligences, including our progress into space and the settlement of Venus at the expense of the sentient aquatic creatures already living there.

Fictional aliens from far and wide flourished during the golden age of science fiction, an episode in mass-market publishing history fuelled in part by the availability of cheap paper made from wood pulp (hence the term pulp fiction). The other prime influence on science fiction was the rapid development, between around 1900 and 1960, of new technologies such as cars, planes, radios, submarines, movies and TV, and of course the first tentative real-life steps towards rocketry and space exploration. Alas, all too many stories, whether in print, film or broadcast media, featured aliens clutching ray guns, their spare tentacles wrapped sinuously around helpless blondes.

A real-life panic

Some alien fictions were more convincing than others, but one in particular felt so true to life it caused mass panic. On 30 October 1938 (Halloween Sunday), Howard Koch and the waywardly brilliant theatrical impresario and soon-to-be filmmaker Orson Welles (1915–85) adapted H.G. Wells's *The War of the Worlds* as a radio play, transmitted as part of Orson Welles's *Mercury Theatre on the Air* series for America's CBS network. The drama included a realistic news broadcast apparently interrupting a big band music show with the following bulletin:

> "It is reported that at 8.50pm a huge, flaming object, believed to be a meteorite, fell on a farm in the neighbourhood of Grovers Mill, New Jersey, 22 miles from Trenton."

Then it was back to the band, except the very credible "news reporter" kept interrupting with ever more alarming accounts of hideousness and destruction at the whim of unearthly tentacled invaders. A sombre voice announced:

> "Ladies and gentlemen, I have just been handed a message that came in from Grovers Mill by telephone. Just one moment please. At least forty people, including six state troopers, lie dead in a field east of the village of Grovers Mill, their bodies burned and distorted beyond all possible recognition."

It's impossible to say just how many listeners thought the bulletin was genuine, and that Martian invaders really had landed in New Jersey. But it certainly frightened many thousands of people among the millions of *Mercury Theatre* listeners that night, especially those who tuned in a few moments too late to hear the show announced as just an exercise in theatrical make-believe.

More sedate visions of Mars

During the 1950s most scientists reconciled themselves to the fact that Mars was not a likely abode for intelligent forms of life. However, they still expected vegetation of some kind, perhaps on the level of lichens clinging to rocks. As the space planners in the US and Russia developed their potent rocket technologies, Mars became a favoured target. A manned exploration seemed possible, in theory at least, for the next generation, sparking popular interest and leading to the publication of several influential books:

▶ **Das Marsprojekt (1952)** In this non-fiction title, German rocket engineer Wernher von Braun (1912–77) outlined how a Mars mission might be possible with 1950s technology. Ten ships, weighing four thousand tons apiece, would make the trip. They'd land using huge glider planes with giant wings capable of grasping the tenuous Martian atmosphere to slow their descent. His ideas were incorporated into a grand vision of space exploration, popularized in family magazines and books. These helped to establish the political groundwork for the founding of a new US government agency, the National Aeronautics and Space Administration (NASA), in 1958.

▶ **The Sands of Mars (1951)** British-born radar scientist, novelist and space enthusiast Arthur C. Clarke wrote this fine science-fiction novel depicting a self-sufficient colony of earthlings breaking away from Mother Earth. Martin Gibson, a visitor to the colony, encounters a secret project aimed at increasing the oxygen content of the Martian atmosphere, and making it breathable, by exploiting plant life. The colonists' ultimate ambition is to set off a thermonuclear reaction on Phobos, one of Mars's two small moons, and turn it into an artificial sun that would burn bright for a thousand years, adding warmth to the planet below.

▶ **The Martian Chronicles (1950)** Ray Bradbury (1920–) used the Red Planet as the setting for one of the most famous science-fiction works of the twentieth century: a poignant short story collection that was rather more about us than "them". Bradbury's human explorers flee an Earth ravaged by war and find that Mars is inhabited by the ghostly remnants of an alien culture. He depicts the carelessness of humans when they enter a new and unknown environment. In one obvious dig at *The War of the Worlds*, the Martians are ravaged by bugs inadvertently brought to their planet by the humans, echoing the unfortunate history of Native Americans' encounters with European settlers. As the story cycle draws to a close, the children of human settlers on Mars stare at their reflections and recognize themselves as what they have become. *They* are now the Martians.

▶ **Stranger in a Strange Land (1961)** Robert Heinlein (1907–88) promoted his hopes for the betterment of humanity by using Martian intelligence as an exemplar in his story about a human born on an expedition to Mars, and stranded there after the death of his parents. Subsequently raised by Martians, and returning to Earth when a second human expedition arrives to find him, Valentine Michael Smith tries to adjust to Earth's bizarre ways, and then to educate humans towards a better and more spiritual way of life, based on what the Martians had taught him. The book's spiritual message fed into the 1960s counterculture, and one of its key practices – of "grokking" or grasping the essence of a vital experience or idea – found its way into the legacy of another major science-fiction hero when "I Grok Spock" t-shirts appeared shortly after *Star Trek* first aired. David Crosby even wrote a song called "Stranger in a Strange Land" for The Byrds. Smith's open-minded and unpossessive attitudes to sex also found a receptive audience in the 1960s.

Cold War screen aliens

As space flight became ever more feasible during the 1950s, Hollywood's film and television industry responded with Mars-related films of varying quality. Plenty of films from this period were little more than creature features in which people ran screaming from monsters in rubber suits, but a handful did have something worthwhile to say about alien life. In 1953, George Pal made a compelling movie version of *The War of the Worlds*, along with a couple of less dramatic but technically impressive space exploration films. While the better types of

"Watch the skies, everywhere! Keep looking. Keep watching the skies!"

Journalist Ned Scott, *The Thing from Another World* (1951)

science-fiction literature explored realms far beyond Mars, it's fair to say that, so far as the general public was concerned, "alien" was more or less synonymous with "Martian" from the time of H.G. Wells to the dawn of the Space Age.

Some fine 1950s science-fiction films included *Invasion of the Body Snatchers* (1956), a paranoid parable of Soviet Communist takeover or US McCarthyist witch-hunting, depending on how audiences chose to interpret it, and *The Day the Earth Stood Still* (1951), in which Michael

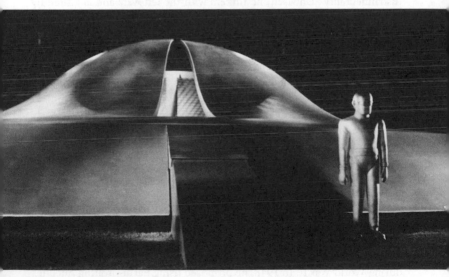

Robotic peacekeeper Gort stands guard over his sleek silver-grey saucer in *The Day the Earth Stood Still* (1951).

Rennie arrives in Washington in a sleek silver-grey saucer and warns the world that it must give up its nuclear weapons, or else be punished by a superior galactic collective of civilizations. His tall robot Gort (also available in sleek silver-grey) is left behind as a police presence. Both these classic movies have been revisited by recent filmmakers, to little benefit.

One movie from that era that should be revisited, but for some reason hasn't, is Fred Wilcox's charmingly chilling interplanetary take on Shakespeare's *The Tempest*. In *Forbidden Planet* (1958), a commander played by Leslie Nielsen (later of *Airplane!* fame) arrives on planet Altair IV to rescue stranded scientist Morbius (Walter Pidgeon). Nielsen and his crew are beguiled by Morbius's daughter, the shapely Anne Francis. Equally impressive in the limbs department is Morbius's intelligent mechanical helper, Robby the Robot, a stunning prop by the standards of his day. Morbius has no wish to leave, because he's exploring the monumental remains of an alien civilization, the Krell. Their machinery still hums in vast underground caverns, and they were quite obviously peaceful and immensely clever, yet there's no sign of them. Where did they go?

As Morbius and his would-be rescuers discover, the Krell had developed technologies so advanced that their mere thoughts could be made manifest. They had become, as Morbius puts it, a "civilization without instrumentality". Unwisely, he tinkers with the leftovers and is visited by the subconscious dark side of his personality, expressed as a destructive near-invisible force field with the vaguest outline of a ghost or a monster. The Krells must have suffered a similar fate, quite accidentally bringing into reality their deepest anxieties as well as their fondest dreams. Not bad for 1950s hokum, and a clear inspiration for many of *Star Trek*'s later adventures. More significantly, *Forbidden Planet* sets up a genuinely alien environment; a source of wonderment and curiosity that is not obviously explicable to us, or to the characters in the film.

Aliens in "realistic" science fiction

Arthur Charles Clarke (1917–2008), the most influential "hard" science-fiction author of the twentieth century, made his mark as a real-life scientist at an early age (he was given the nickname "Scientific Sid" at school) and helped pioneer radar landing systems for British bombers during World War II. In 1948, he proposed using Earth-orbiting satellites as the basis for a global communications industry, complete with all the equations for orbits and radio frequencies.

He spent the rest of his life advocating human expansion into the surrounding solar system, as a writer of both non-fiction and science fiction. Two books in particular, *The Exploration of Space* (1951) and *The Exploration of the Moon* (1954), each with a lively mix of populist language and strict scientific accuracy, were hugely influential among scientists and aerospace engineers preparing for the realities of rocket travel. Clarke also wrote a series of famous novels exploring our place in the cosmos. For him, alien intelligence was never merely a science-fiction trope. It was an abiding scientific and philosophical passion.

In *Childhood's End* (1953) an alien fleet visits Earth to prepare us for our final destiny. The last human generation is absorbed into a greater cosmic consciousness, while the merely physical world is annihilated. The human survey team of *Rendezvous with Rama* (1973) discovers a colossal alien ship, empty but apparently capable of housing thousands of aliens. The craft plunges through the sun, collecting energy along the way. It departs as quietly as it came, completely ignoring its temporary human visitors and their misguided attempts to save it from crashing into the sun. Perhaps most famously, a short story, "The Sentinel", originally written in 1948 for a BBC competition (it didn't win), was the basis for Clarke's collaboration with filmmaker Stanley Kubrick on the most durable and famous science-fiction film of all time: *2001: A Space Odyssey*.

Released in 1968, *2001* is slow-paced and narratively ambiguous, yet never less than ravishing to look at. It portrays our initial encounters with an alien intelligence, first on the moon, and later in orbit around Jupiter. The film's scientific credentials were impeccable, and the alien encounters showed nothing that couldn't potentially happen. No bug-eyed creatures stepped out of flying saucers spouting American-English, and no one carried ray guns. In fact, Kubrick and his screenwriting collaborator, Clarke, elected not to show the aliens at all. Instead, the unseen entities were represented by their eerie technology in the form of simple black slabs, or monoliths. *2001* looks as fresh today as it did when it was first released.

The film also explores themes of artificial intelligence. On board the human spaceship *Discovery*, supercomputer HAL9000 rules the roost until he gets ideas above his station and has to be dismantled. With their crisp geometry, the alien slabs can be interpreted as artefacts with mysterious powers. HAL and the monoliths show that there needn't be any intrinsic difference between intelligent life and intelligent machines. Certainly we are already at a point in human history where we can predict ever closer unions between our biological selves and our technologies. Highly advanced aliens may have blurred the distinction to the point where it vanishes.

Ten years later, Kubrick's visual effects collaborator Douglas Trumbull created an awe-inspiring saucer for Steven Spielberg's *Close Encounters of the Third Kind* (1977). This was the next major science-fiction film in which alien contact was treated as a serious possibility, although Spielberg elected to show his creatures as grey, skinny, child-like humanoids, basing his designs on how countless UFO enthusiasts imagined benevolent extraterrestrials should look. Rather than landing without invitation, they signal their friendly intent, and a formal welcoming party awaits them in a remote desert location, even though it's gate-crashed by civilians who've been acutely receptive to the aliens' telepathic invitations. As in *2001*, it's assumed that government officials will be nervous about telling the rest of the world about the encounter for fear of causing unrest or panic. But unlike so many flying saucer movies, *Close Encounters* insists the aliens will be friendly, and that, at some point, the rest of the human world will learn of their arrival.

Ranking alongside *Close Encounters* is *Contact* (1997), in which Jodie Foster plays Ellie Arroway, a radio astronomer on the lookout for alien signals. After years of methodically analysing radio noise from space, she and her team finally detect a repeating pattern of prime numbers. The signal turns out to be a set of instructions for building a giant machine that will facilitate contact between humans and a distant alien intelligence. The machine is constructed, but only after tremendous political and philosophical uncertainty on the part of the earthly authorities. Arroway gets to ride the device, a small pod suspended in vast wheel-shaped energy transmitters, and she experiences a mind-warping eighteen-hour voyage across the universe.

On "returning" to Earth, Arroway is stunned when her disappointed colleagues explain that her pod simply fell from its cradle to the ground, apparently going nowhere. Only the fact that Arroway's video camera has recorded hours, rather than seconds, of meaningless static supports her version of events. Scientifically and philosophically fascinating, it's no surprise that this plausible film is based on a novel of the same name by noted planetary scientist Carl Sagan (of whom, much more later).

> **"This isn't a person-to-person call. You can't possibly think that a civilization sending this kind of message would intend it just for Americans!"**
>
> Radio astronomer Ellie Arroway, *Contact* (1997)

With its careful exploration of ideas concerning both science and faith, *Contact* is the benchmark for movies about alien intelligence.

The idea of instructions beamed to us from outer space was also explored by renowned British cosmologist Fred Hoyle, in the 1961 BBC television drama series (and subsequent novel) *A for Andromeda*. A radio

The ultimate alien encounter movie

Four million years ago, our ape-man ancestors struggle for survival in an arid landscape. There are frequent disputes for possession of a muddy waterhole. At night the apes huddle in a cave while predators prowl outside. So begins the 1968 movie *2001: A Space Odyssey*.

One morning a mysterious black slab appears outside the cave. Then it vanishes, but some while later, an ape leader picks up a bone from a stray carcass and, half remembering the slab, suddenly understands the bone's potential as a weapon. He kills other animals for food, and then disposes of a rival leader to win the precious waterhole. Triumphant, and assured of survival, he throws his bone into the air. Instantly it becomes a spaceship in the year 2001.

American astronauts find what looks like the same monolith on the moon, but they have no idea what it means. On seeing sunlight for the first time in millions of years, it sends a radio signal towards Jupiter. A spaceship is sent to investigate. The crewmen in the ship are not told the reason for their flight, but HAL 9000, the sentient on-board computer, has been informed. A conflict arises when HAL decides that his human companions are less important to the mission than he is. Only one astronaut survives HAL's destructive efforts.

After disconnecting HAL, the lone astronaut is sucked into a vortex of distorted space and time. He ages swiftly in a faux-antique hotel room, perhaps drawn from his imagination. In the finale, he is transmuted into a large-eyed baby drifting towards Earth in a glowing bubble. What happens next, and what the baby means, is up to the audience to decide, although Arthur C. Clarke's novelized version, released in the film's wake, goes some way towards offering an explanation (as does the sequel, *2010: Odyssey Two*, published in 1982).

message received from within the Andromeda nebula contains instructions for building a supercomputer. When constructed, it lays down yet more instructions, this time for creating some kind of living biology. As the investigating scientists bicker over control, some of them sense a tremendous opportunity for technological and military advancement, while others want to shut down the project before unpredictable disasters ensue. Cutting a long and intriguing story short, the computer directs the cloning of an alien-enhanced young assistant (played by Julie Christie), who is then put to work on secret defence projects. Greed threatens to blind humanity to the risk of a subtle form of alien invasion. It's an interesting plot, not least for its exploration of how we might react when confronted by technologies more powerful than our own.

Metaphors for human anxiety in the modern age

A slew of more recent movies, from *Independence Day* (1996) to *Battle: Los Angeles* (2011), have revisited the wham-bang business of alien invasions, in the process revealing something about human concerns. A superior example is Spielberg's *War of the Worlds* (2005), a stark and sober retelling of the familiar story, with a nihilistic post-9/11 outlook, in which Tom Cruise tries to save his family from Martian machines that have been buried in the Earth for aeons, and are now bursting through the crustal plates under New Jersey to complete their terrible takeover mission. When the invaders finally succumb to simple infections, there is no jubilant victory for the surviving humans. The great catastrophe that has visited Earth ceases as suddenly and pointlessly as it began.

Just as mid-century science-fiction movies often expressed Cold War paranoia, responded to people's fears about technology running amok or questioned what it is to be human, so contemporary movie and TV aliens continue to reflect human hopes and anxieties. One such telling theme can be found in the TV series *Battlestar Galactica* (2004–09), an ambitious and dark reworking of a juvenile 1970s series designed to cash in on the success of *Star Wars* (1977). In the new version, robotic Cylon aliens are war machines originally created by humans, and now violently seeking independence. The Cylons gain the upper hand, along the way discovering how to mimic humans, even down to their blood, sweat and tears. Chilling scenes of interrogation and torture pose questions of moral

Alien archaeology

If real aliens do show up one day, we may be more than a little embarrassed to exhibit our film archives – a theme explored in Arthur C. Clarke's short story, "The History Lesson" (1949). Far in the future, a Venusian science team arrives on Earth and discovers a few tantalizing fragments of the technological race that must once have thrived there, but has long since vanished. One object that intrigues them is "a flat metal container holding a great length of transparent plastic material, perforated at the edges". After further study, they discern faded images on the strip and work out how to display them in something like the correct sequence. The Venusians interpret the images as "a record of life as it was on this planet at the height of its civilization", and watch, spellbound, as a strange creature becomes "involved in an incredibly violent conflict with another, slightly different creature". What would have been, to our eyes, an innocent bit of cartoon tomfoolery leaves the Venusians reeling with shock.

authority in times of crisis, with clear references to the modern "war on terror" Cleverly dressed up in the populist context of spaceships and laser guns, *Battlestar Galactica* is serious entertainment, but as usual, it's not really about the aliens.

Neither is that ultimate space horror movie, *Alien* (1979), Ridley Scott's masterpiece of grunge-tech misery on a space cargo ship infested with a carnivorous creature with slobbering jaws and acid for blood. The extraterrestrial monster lurking in the shadows is a theme of countless movies, ever since *The Thing from Another World* (1951) first menaced a terrified team of scientists at a lonely Arctic research base. We can only hope that human technology never achieves the *Forbidden Planet*'s craft of turning thoughts into reality, in case we inadvertently conjure to life Scott's terrifying beast from the dark recesses of our own nightmares.

Despite the breadth of filmmaking styles and budgets over the years, most fictional aliens are only there to give humans something to react against, whether by shooting them or cuddling them. We've even fallen in love with them and had their babies, as in John Carpenter's *Starman* (1984), where an alien disguised as Jeff Bridges gives charismatic comfort to Karen Allen's lonely young widow. But mainly, we shoot them.

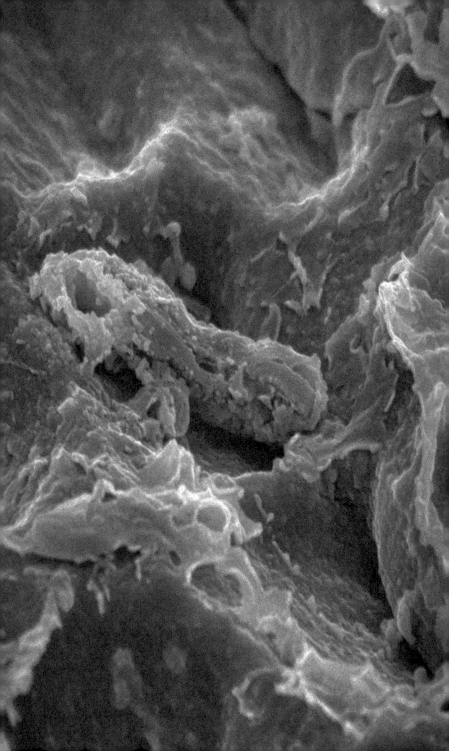

The origins
of life

There's no doubt that the cosmos is awash with life's basic building blocks, but right now there's only one world on which we can study how planets arise, and try to guess how life gets started on them.

Where to begin in our genuine scientific hunt for extraterrestrials? Everything we know about life comes from what we've found on Earth. Most of what we know about the potential origins of life bearing planets comes from the same place. Just because the Earth exists is no guarantee that other regions of space must be peppered with similarly life-infested worlds. On the other hand, all the physical forces and chemical ingredients that gave rise to the Earth are commonplace throughout the galaxy. So let's start by looking at our own planet, and see where this gets us.

Origins of the Earth

Earth is a product of colossal explosive events that occurred in the Milky Way galaxy long before our sun, or any of the planets in our solar system, came into being. Around five billion years ago, a vast cloud of dust and gas drifted through the galaxy. These thin, widely scattered wisps originated from supernovae, the explosive deaths of stars from an earlier generation. The cloud, called a **nebula**, was gently pulled around by gravitational forces from surviving stars in the surrounding galaxy. Most of the atoms in the nebula were hydrogen or helium, but all the other chemical elements were present in smaller amounts.

Somewhere within the nebula, a seemingly insignificant clump of matter coalesced. It gained mass until it was capable of exerting its own gravitational force, extremely weak, yet just strong enough to attract nearby atoms and molecules. This mass, known by scientists as a protostar, grew to the point where its gravity became strong enough to pull in more material from ever greater distances: tens, then thousands, and eventually many millions of kilometres. Dust and gas from the surrounding nebula became concentrated into a spinning plate-shaped structure known as an **accretion disc**, with the protostar – our young sun – at its centre.

Even as the new sun gathered mass, some of the chemical debris in the accretion disc began to cluster in miniature, and much cooler, versions of the protostar process. Small worlds, or " began to form. When these collided with each other and their wreckage recombined, larger structures emerged and consolidated into the planets and moons that exist today. What we don't know is how many planetesimals existed in the solar system's early days, and how many were shattered by chaotic collisions. The asteroid belt that drifts between the orbits of Mars and Jupiter is testament to at least one very violent destruction of a sizeable planetesimal. Some scientists speculate it could be the result of an object about half the size of the Earth's moon coming to grief somehow.

One planetesimal became our Earth. It began life around 4.6 billion years ago as a hot, chaotic sphere of molten metals and silicates. Over time the various materials began to divide according to density, with the heaviest metals, such as iron, sinking into the Earth's core, and the lighter materials, such as the silicates, floating upwards towards the planet's surface. After approximately 100 million years, the outer surface cooled sufficiently for a thin rocky shell (the crust) to form.

As the Earth's crust cooled and hardened, weak spots allowed molten material from beneath to burst onto the surface through volcanoes. These eruptions also hurled out massive amounts of carbon dioxide, ammonia and methane, creating the early atmosphere (the precise mix of gases is subject to debate, as we'll see later). In addition, the volcanoes emitted water in the form of super-hot steam, which condensed into clouds, then cooled still further to fall as rain. So emerged the first shallow seas of liquid water on the Earth's surface.

Origins of the moon

We once thought that our moon must have been a planetesimal captured by the Earth's greater gravity field. Now a much more violent theory is favoured. When the Earth was less than 100 million years old, and while its crust had barely begun to harden, it collided with another newly formed object almost as large as Mars. The Earth's delicate new crust was

Pockmarked: the craters covering the moon are clues to its violent past.

extensively reshaped, while the Mars-sized impactor world was smashed altogether. Its debris then orbited the injured Earth, intermingled with a considerable proportion of wreckage from the Earth's own crust that was hurled into space by the catastrophe. Gradually this ring of rubble coalesced into a new world, the moon. Mutual tidal forces between the Earth and the moon have slowed the moon's orbit around its own axis so that it keeps one face towards the Earth at all times. People talk of the "dark side" of the moon, but it is more accurate to call it the "far side". It receives just as much sunlight as the familiar face we always see.

In full daylight the moon's surface can reach blistering temperatures of 100°C, plunging at night (which lasts fourteen Earth days) to -180°C. The moon is pockmarked with craters from countless asteroid impacts, as if at some point it was the victim of an especially violent episode. In fact, most of its craters were accumulated only gradually, throughout 4.5 billion years of lunar history. In those same 4.5 billion years, Earth has also been smashed into countless times.

Sorting the solar system's worlds

The first generation of stars, created at the dawn of the galaxy, consisted of nothing but hydrogen and helium, the simplest units of matter. All the other elements – the basic chemicals which create the richly variegated structures of our world today – had to be manufactured in long-vanished suns as by-products of nuclear fusion. Those suns then had to scatter their products in the vast clouds from which our modern solar system was born.

When the solar system formed, the sun's radiation boiled away the lighter hydrogen molecules in its surrounding accretion disc. The regions closest to the star became richer in heavier elements like iron and silicon, while the outer, colder regions were able to hold on to hydrogen. This division by density was possible only because the system's original gas cloud contained such a variety of substances.

This solar-blast grading in the solar system's building materials, from heavy to lightweight, explains why the inner planets, Mercury, Venus, Earth and Mars, are made principally of silicate rocky crusts with hearts of molten iron, while Jupiter, Saturn, Uranus and Neptune are gaseous giants with no rocky crusts, little if any iron content, and cores made of hydrogen, compressed so tightly by gravity that it takes on a weird metallic form unlike anything that we could recreate on Earth. But it's still hydrogen.

There was, however, a particularly unstable period of Earth's evolution, from about 4 billion to 3.8 billion years ago, known as the **Late Heavy Bombardment**. A vast amount of debris left over from the solar system's formation drifted through nearby space. Rogue planetesimals and rock fragments frequently collided with the Earth, repeatedly smashing the crust. Subsequent wind, rain and erosion has smoothed away most traces of these catastrophes, but we only have to glance at the countless craters on the airless moon to see what the young Earth suffered. With destruction, however, came the possibility of creation. Some asteroids and comets falling on the Earth may have contained chemical compounds that contributed to the emergence of life.

Einstein and the laws of physics

Earth's origins make a good case study as we search for similar planets beyond our solar system. Leaving aside the question of life just for the moment, our particular ball of rock seems to be a fairly typical product of planetary formation. These processes are governed by the laws of physics, which **Albert Einstein** (1879–1955) insisted must be the same throughout the observable universe. Certain special values in the natural world are constant everywhere. For instance, the speed of light traversing a vacuum is the same in all places, whether in our solar system or in the depths of a far-off galaxy.

A hundred years later, Einstein's famous predictions about matter, energy and relativity hold up almost perfectly. Some physicists are tinkering with the idea that light might not always have travelled at the same velocity throughout the entire 13.75 billion-year history of the universe, and that some other fine-tuned constants, such as the strength of gravity or the forces that bind atoms together, might perhaps be changing by infinitesimal degrees as the universe ages and expands. These are speculations on the wilder frontiers of the theoretical landscape, and have no great relevance over the petty span of cosmic time that encompasses the human era.

To cut a long story short, we can be confident that all regions in the observable universe are subject to the same laws of physics. Our sun works by nuclear fusion, and all other stars that are similar in mass, luminosity and age will be born, live and die in the same way. Light from even the most distant stars and galaxies reinforces our conviction that the same laws of nature apply everywhere. In principle at least, all types of physical event that have taken place in our galactic neck of the woods – such as the creation of planets and the emergence of life – could occur elsewhere in the universe too.

The Mediocrity Principle

According to the **Mediocrity Principle**, if the Earth came into existence orbiting a commonly available type of star, it's reasonable to assume that other solar systems can come into existence around other stars, so long as the local conditions happen to be similarly appropriate. Furthermore, the chemicals that make up Earth, and everything upon it, are not exclusive to us. All atomic elements, and at least some of the organic molecules essential for life, are commonly available throughout the universe.

Essentially, then, the physical and chemical conditions on our Earth are mediocre. In this context, mediocrity doesn't mean "pertaining to the

The composition of Earth's crust	
Oxygen	46.6%
Silicon	27.7%
Aluminium	8.1%
Iron	5%
Calcium	3.6%
Sodium	2.8%
Potassium	2.6%
Magnesium	2.1%
Other	1.5%

condition of rubbish". The phrase comes from statisticians' use of the word median, meaning "average". Now the challenge is to find out if the Mediocrity Principle applies to life as well as planets.

Life takes hold

Although we can't yet say exactly how, or where, the first organisms emerged, they were almost certainly **cyanobacteria**. Comprising single-celled bacteria amassed in vast colonies, these blue-green algae used sunlight (photosynthesis) to derive their food and energy, in much the same way that plants do. Mineral rock records show that the first life forms emerged with startling speed soon after the Earth's crust solidified. By around 3.8 billion years ago, the oceans supported a great mass of algae. Since there were no more complex organisms around to eat them, they lived and died in relative tranquillity across countless generations.

The main threat to the algae's survival was solar radiation, not predation. Pigments in their cells protected them against ultraviolet rays while letting in beneficial wavelengths of light for photosynthesis, enabling the algae to break apart water molecules from the surrounding ocean and use the hydrogen in their internal chemistry. The hydrogen was combined with carbon extracted from the atmosphere's carbon dioxide, to manufacture hydrocarbons essential for their livelihoods. The one thing they *didn't* want was too much pure oxygen, since they were anaerobic, or "oxygen-hating", microorganisms.

The product of these first biological entities was a form of atmospheric pollution as destructive as any ecological horror we may fear today. After a billion years and more of global domination, the blue-green algae gradually became a victim of their own success, pumping out vast quantities of oxygen that changed Earth's atmosphere into a lethal mix they could no longer tolerate. This was the origin of the atmosphere that we breathe today. Oxygen-based reactions create more energetic chemical exchanges than non-oxygenated ones. This new bias in the environment fed an additional kick of energy into those organisms capable of adapting to the altered atmosphere.

Oxygen-breathing life forms began to lead more complex and eventful lives than their cyanobacterial forebears. When some enterprising little bugs developed the technique of absorbing others for nourishment, the innocent age of the cyanobacteria came to an end, and evolution really hit its stride. Some bugs merged into cooperative populations to become complex multicellular creatures. At that point, around 700 million years ago, the ever-increasing "technology" of living things – swimming, crawling, sensing, hunting and avoiding being caught – paved the way for a surge of evolution that continues to this day.

How evolution works

Charles Darwin (1809–82) didn't invent evolution. What he did, and so brilliantly, was describe how it works. The seemingly callous truths about the natural world generate heated argument because of their dependence on dispassionate chance rather than deliberate godly design. Yet the fact remains that evolution is the best way to explain why living things are so superbly well-adapted for the lives they lead. Along with pain and death, evolution has delivered beauty and joy, curiosity, wonderment, consciousness and love – not just among humans, but probably among many other animal species as well. Okay, so the animals do like to eat each other now and then, or have wars with sharp implements and bombs. Big deal. Evolution is rough and tough, but it's not all bad.

> **"I see no good reason why the views given in this volume should shock the religious feelings of anyone."**
> Charles Darwin, *On the Origin of Species By Means of Natural Selection* (1859)

As identified by Darwin in *On the Origin of Species by Means of Natural Selection* (1859), the mechanisms of evolution are stark and simple. All living things replicate themselves by reproducing. In so doing, they transmit certain physical and behavioural qualities that enable their offspring to survive in the environment in which they live. If something in that environment changes – for instance, a warm, moist climate gradually turns cold and dry, or new and dangerous predators arrive on the scene – then the species must adapt to these changed circumstances or die out.

At first glance, all of the offspring of a given plant or animal seem to look more or less like their parents, but they are never exact copies. Some may have slight but significant differences in body shape, external colouring or instinctive behaviour that can affect their ability to survive, for better or for worse. Beneficial differences continue to be passed down through the generations, but poor ones are eventually weeded out because they do not aid survival, and the creatures or plants exhibiting them fail to thrive. These unfortunate losers in the evolutionary trials of life produce fewer offspring, and eventually become extinct.

Evolution's finer details

Darwin showed that such differences, although subtle from one generation to the next, can become markedly pronounced over many generations, thereby creating new species. But he was uncertain exactly how this
mechanism worked. Today we know that it happens at a molecular
level, within the DNA that determines the growth and development of
all organisms. External influences, such as radiation from the sun, can
mutate DNA. Occasionally this may nudge it in a potentially survivable
direction, but usually the effect is harmful.

Sexual reproduction is a more valuable contributor to genetic change
because it routinely shuffles the genes encoded within DNA, so that no offspring organism ever has exactly the same genetic sequence as either of its
parents. Reshuffles can go wrong. Inherited diseases or predispositions are
a product of harmful rearrangements of genes. Most of the time, however,
the new DNA created from parental contributions functions properly, and
the offspring gets to try its chances in the lottery of life and reproduction.

As modern evolutionary champion Richard Dawkins points out, "The
secrets of evolution are death and time." The biggest factor shaping
evolution is the constant threat of annihilation for the individual, and
extinction for a species. The surrounding environment can change with
spectacular suddenness, for example when once reliable seas dry out, or
deserts get flooded. Organisms whose major genetic characteristics have
survived unchallenged for tens of millions of years can be wiped out
in a few generations. Of all the species that have existed since the first
emergence of life, almost all are extinct, or else mutated into new forms.
Come to think of it, of all the gods that human beings have ever believed
in, almost all of them have fallen into disuse, because the cultures that
gave rise to them underwent their own changes and extinctions. Clearly,
gods face survival pressures too.

In 2008, biologist Lee Grismer from La Sierra University, California,
found a hitherto unknown species of gecko in the forests of an island
off northwestern Malaysia. He also discovered similar ones in nearby
limestone caves, where they were seeking to escape their main predator,
the pit viper. But the geckos in the cave were subtly different from their
forest-dwelling equivalents. To live and hunt in a cave, they had changed
their body shapes, and had flatter heads, longer limbs and slighter builds.
If Grismer's ideas are correct, we are witnessing **speciation** in action: the
moment – measured in years and decades, rather than centuries or millennia – when one animal species starts to give rise to another.

Evolution in action

If extraterrestrial life is discovered, evolutionary forces will have shaped that life, from its earliest and stinkiest beginnings in some alien pond to complex creatures in starships, or whatever. Even if an alien entity turns out to be technological rather than biological (an intelligent robot, perhaps, or a sentient spaceship), then that technology will have been built by – or at least, preceded by – something living, which, in turn, will have evolved.

Natural selection usually operates over vast time spans, but sometimes we can witness it happening over just a few years. The peppered moth *Biston betularia* is typically a grey-white creature with small black spots on its wings, disguising it as it sits on the bark of certain kinds of birch trees. In the mid-nineteenth century, English naturalists noticed that the bright bark of the trees on which the moths rested had become dulled by industrial pollution. Birds were now finding and eating more peppered moths than usual, but missed the very occasional darker ones.

In 1850 there were twenty light-coloured peppered moths for each dark one. Just fifty years later, as the birch bark blackened under the remorseless onslaught of the industrial age, there was one light peppered moth to twenty dark ones. A seemingly insignificant mutation in the colouring of the moth's wings, requiring just a minor adjustment in the moth's DNA, made a fantastic difference to its chances for survival. Interestingly, the story continues. Since the 1950s, the Clean Air Act has prevented British factories from belching out filthy soot. In the cleaner, post-industrial landscape of today, the birch trees are not so dirty, and the dark peppered moths, for a while so successful at hiding in the grime, have lost their advantage and been picked off, leaving lighter moths behind. In 2009, UK-based charity Butterfly Conservation reported the moths "making a big swing back to their original colour", and appealed for new studies to assess the dark–light distributions. They determined that the darker moths had "undergone a major decline of 61 percent since the 1960s".

A light-coloured peppered moth visible on the dark bark of a birch tree.

Evolution drives life, even when sentient creatures, such as human beings or advanced extraterrestrials, invent technology in a bid to subvert it. As a case in point, the emergence of farming was once thought to be responsible for slowing down natural selection in humans. By storing up food through the winter months, we seemingly protected everyone alike, thus counteracting nature's merciless tendency to weed out the weaklings. It turns out, however, that many humans descended from agricultural ancestors that adapted to drinking milk from cows and goats. The guts of their ape-man ancestors would have rebelled at this indigestible stuff. Far from halting our evolution, technology simply drives it in different directions.

As we turn our attention to the search for alien intelligences, this milky story gives us a telling clue about the relationship between biology and technology. Far from existing in separate realms, they are facets of the same evolutionary process. Ancient cultures devised farms, and the availability of milk from those farms caused our digestive systems to adapt to the new foodstuff. Modern technology, from the Internet to genetic engineering, must also create some kind of biological feedback response – even if we won't be able to tell, for many generations to come, what direction that response might take. Similarly, when we hunt for traces of other life in the universe, we shouldn't necessarily expect any distinction between a purely biological organism, or one shaped in large part by its own technologies.

Itsaq: the oldest traces of life on Earth

There is a rock formation in southwestern Greenland known as **Itsaq**, or "Ancient Thing". The rocks were belched up by volcanoes nearly four billion years ago, when the Earth was barely a ninth of its present age. Since then, more than 99.9 percent of the planet's entire surface has been remelted, reshaped and generally transformed. Itsaq (or the Itsaq Gneiss Complex in geologists' parlance) is an exceptional snapshot of a time when the world was young.

Much of the ground we stand on was once covered by the muddy ooze of an ocean floor, which over time was pushed above sea level by the Earth's ceaseless geological forces. The layers of soft sediment were compressed and hardened into dense rocks, such as those now found at Itsaq. They appear to contain traces of early life – not fossils of the organisms themselves, but the subtle remains of their chemical ghosts.

In the summer of 1996, a team from the Scripps Institution of Oceanography in California, led by geochemist Gustaf Arrhenius, examined these rocks. They contain carbon compounds, such as apatite, that cannot instantly be dismissed as natural mineral products. **Apatite** is a common enough compound of calcium, carbon, oxygen, phosphorus and fluorine found in many rock formations. But it's also associated with life; indeed, the outer enamel of our teeth consists largely of apatite.

In a paper for *Nature* magazine, the Scripps team analysed carbon isotopes in tiny grains of apatite within the Itsaq rocks, and concluded that the grains are probably biological in origin. Their claim, which no one has yet refuted, is that life existed on Earth at least 3.85 billion years ago – and if it existed then, it must have got going even earlier. Push the starting date much further back, and the Earth's crust would scarcely have been solid yet.

Itsaq's subtle clues are open to interpretation, because they lack the obvious structural complexity that we might expect in a "fossil". Samples taken from slightly younger rocks near Strelley Pool, a watering hole in the remote Pilbara region of Western Australia, are more convincing. These rocks definitely contain microfossils: not just chemical traces but firm evidence for biological cells. The 3.4 billion-year-old rocks at Strelley Pool were once on the shorelines of an ancient sea. As we push at the limits of time, there's a growing suspicion that life didn't exactly have to struggle to find a foothold on the young Earth.

Telling one carbon atom from another

Chemically, all carbon atoms in the universe are exactly the same. They have six electrons buzzing like gnats in their outer orbital shells, and six protons packed into their dense nuclei, and they employ the same tricks during chemical reactions. That said, some carbon atoms are slightly heavier than others. Their various isotopes are defined by different numbers of neutrons in the nucleus. Neutrons are quite weighty, but are not involved in binding with other atoms, and carry no electric charge. Biochemistry favours the lightest isotope, carbon-12, where six protons are matched in the nucleus by six neutrons. Carbon-13 packs an extra neutron, making it heavier. On Earth, there is a distinct bias towards carbon-12, because this lighter isotope is a little swifter and more mobile in its reactions.

What does all this mean? Essentially, if we encounter an organic molecule packed with a random mix of carbon isotopes, chances are the molecule was never part of anything alive. On the other hand, if the molecule exhibits a marked bias towards the lighter carbon isotopes, then it is likely to be the product of biology. The carbon atoms in the Itsaq apatite grains were almost entirely carbon-12. Any similarly distinct bias towards lighter carbon atoms in extraterrestrial organic materials would be cause for curiosity.

How did life begin on Earth?

Darwin did not pretend to have any theories about the origins of life (**abiogenesis**). By temperament and scientific training, he rejected any possibility that it might have arisen spontaneously in some divine puff of magic. In 1871, he wrote an informal letter to his close friend, naturalist Joseph Hooker, in which he speculated about life's origins "in some warm little pond with all sorts of ammonia and phosphoric salts – light, heat, electricity etc – present", and suggested that "a protein compound was chemically formed, ready to undergo still more complex changes".

Some 140 years later, we have yet to improve on that prescient speculation. We know a great deal about DNA and the fantastically complex processes of biochemistry, but we still don't understand how those wonders arose from the obviously far less complicated scraps of chemistry knocking around on Earth at the dawn of its history. If we could solve that riddle, we'd have a better chance of guessing how easy (or difficult) it might be for life to spring up on other worlds.

In the early 1920s, Russian biochemist Aleksandr Oparin (1894–1980) theorized that precursor structures of surprising orderliness might assemble, purely by chance, from simple non-living organic compounds. He imagined droplets of chemicals, with minor variations of oiliness or waxiness creating internal layers, and even distinct boundaries, or primitive membranes, separating each droplet from its environment. These precursor cell-like structures could exchange chemicals with the outside world through their membranes, yet still remain distinct droplets. In 1929, renowned British biologist and geneticist J.B.S. Haldane (1892–1964) speculated how the first organic building blocks might have emerged from the reactions of methane, ammonia and water on the early Earth when exposed to ultraviolet solar radiation. Haldane talked of a "prebiotic broth, or primordial soup".

The Miller–Urey experiment

The soup thing really caught on, and in 1952, Stanley Miller, a University of Chicago graduate student, worked with his supervisor Harold Urey on what's now known as the Miller–Urey experiment. The aim was to cook up some of that soup by simulating the conditions that might have prevailed on the early Earth.

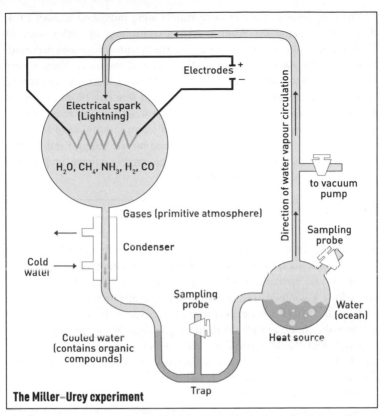

The Miller–Urey experiment

This famous experiment conjured life's building blocks from simple ingredients – but bricks don't guarantee a house.

Miller filled a sterile system of glass flasks and tubes with a mixture of methane (CH_4), hydrogen (H_2), ammonia (NH_3) and water (H_2O), a delightful brew of the kind that he and many other scientists at the time imagined must have belched out of Earth's early volcanoes. The containers were sealed to prevent any contamination from outside this little glass universe. Then the rig was heated so that the water turned to vapour, on the reasonable assumption that the young Earth was pretty hot. Electrical sparks were discharged through one of the flasks to simulate lightning, under the similar assumption that it would have flashed on early Earth as well. Cycles of heating and cooling ensured that condensation of the ingredients would be stimulated just as often as evaporation.

When the results of Miller's experiment were published in May 1953, they caused a sensation. After a week of continuous operation with no further intervention, tremendous and startling transformations occurred. Brown tar-like stains appeared on the inside of the glass. They turned out to be a diverse mix of organic compounds, including an impressive range of amino acids. There was a sense that one day soon, life would be generated in a laboratory from scratch. It was then, and remains now, one of the most exciting moments in all of science. But it was a long way off from replicating what actually happened on the young Earth. In the half-century since those brown stains appeared, we have revised our understanding of the early atmosphere.

A closer look at Miller–Urey

One problem is that pure, chemically unattached hydrogen atoms probably wouldn't have existed on Earth, at least not for long. Hydrogen is so light, it simply escapes into space unless it's bound up in other molecules. Ammonia and methane, two key ingredients of the Miller–Urey scenario, are no longer assumed to have been so widely available either. Leaving that aside, all the Miller–Urey experiment really achieves is to prove how easy it is to make amino acids from simple ingredients in the environment. It's so easy, in fact, that they are often found in meteorites and interstellar dust clouds. Life may rely on amino acids, but on their own, they are merely the simplest bricks, and not the house of life itself.

Today we understand that proteins – which Charles Darwin speculated might "undergo still more complex changes" on the way towards life – are not random chemicals. Shake a flask of amino acids until the cows come home, and nothing will induce them to become proteins, let alone cows. Other proteins, DNA and RNA in particular, need to mediate the assembly processes, controlling how simple amino acids string together to make up the complex and extremely specific information that defines every protein. The sequences are not random. What's more, all amino acids within living things exhibit **chirality**, a particular rotational bias in their molecular structure that helps them link together. Nothing in the Miller–Urey experiment leads us towards this kind of complexity, nor sifts amino acids of the correct chirality from those with an opposite twist. Another ingredient is needed that cannot be sealed in a jar or buzzed with lightning. That ingredient is the most important part of biology, and it can be summed up in one word: information.

DNA: the double helix

Deoxyribonucleic acid (DNA) is often described as the blueprint for life. This is a bit misleading because a blueprint shows us what something's going to look like in advance. DNA is best thought of as an extremely detailed recipe in the form of a biochemical computer program. The list of ingredients in a recipe, no matter how long or detailed, provides no obvious clue as to the final outcome, but when all the necessary processes happen in the correct order, the "meal" emerges in all its splendour. The key to DNA's power as an information system lies in its molecular structure. No matter how long and complex the embedded biological information might be, all of it is encoded using four very simple information units.

DNA is constructed like a surreal spiral staircase whose two handrails twist around each other in opposite directions:

▶ The rungs of the ladder are constructed from four simple compounds, or bases: cytosine (C), guanine (G), adenine (A) and thymine (T).

▶ In any given rung of the ladder, cytosine always bonds with guanine (C–G), or adenine with thymine (A–T), making base pairs.

▶ A base attached to a sugar molecule and a phosphate molecule comprises a nucleotide, the essential structural units from which long strands of DNA are assembled.

DNA's most important quality is that the two intertwined spirals can split completely apart, right down the middle of all the base pairs. Then each half can generate a copy of its missing opposite half, reconfiguring the complete DNA as a distinct and exact copy of the original.

DNA's coding system

The ordering of base pairs in very long sequences defines the information required for growing and maintaining an organism, and for passing its characteristics down to the next generation during cell division. Specific sequences of base pairs along the DNA's length are genes, the instructions that define particular physical and behavioural characteristics of an organism. Each gene controls the manufacture of a single protein. Human DNA has three billion base pairs encoding approximately 25,000 genes. A strand of our DNA is one-tenth of one-billionth of a centimetre wide, but would stretch across nearly three metres if uncoiled into a straight line.

The DNA shuffle

Now let's talk about sex. During a process called meiosis, cells divide to produce gametes, male sperm or female eggs. The DNA replicates as in any cell division, but only half the information content ends up in each parent's gametes. Chucking away half your DNA before passing it on to your beloved offspring might seem quite literally counterproductive, but if all organisms were born with exactly the same DNA as their parents, the slightest adversity affecting one of them would doom all of them. **Gene shuffling** is a better strategy, ensuring that each of your children ends up with brand new DNA based on contributions from both parents, but not exactly the same as either parent's original DNA. This is how genetic diversity is assured. Gene shuffling (and sex, for that matter) may have evolved to outsmart viruses and other infectious agents by constantly staying one step ahead of them.

> **"We are survival machines – robot vehicles blindly programmed to preserve the selfish molecules known as genes."**
>
> Richard Dawkins, *The Selfish Gene* (1976)

Now we come to the ultimate chicken-and-egg riddle. The vastly complicated workings of DNA are dependent on an army of protein facilitators in a cell, but those proteins can't exist without DNA and genes to dictate their structures. At the dawn of life, what came first? The instructions that make and control the proteins, or the proteins required for encoding and enacting the instructions? Are genes, DNA and proteins the only route to biology, or could alien life forms operate according to some radically different principle? We'll return to this question later on (see p.44).

Life from clay?

In the mid-1960s, Graham Cairns-Smith, an organic chemist at the University of Glasgow, suggested that a potential catalyst (a substance that remains unchanged while facilitating chemical reactions) for pre-biological complexity was common in the Earth's environment long before any actual life emerged. Cairns-Smith focused on one catalyst in particular that is literally as common as mud. Upon this happy platform, random amino acids and other simple organic compounds convened by chance, and were gradually encouraged to assemble into something approaching life.

Cairns-Smith isn't interested in ancient oceans or primordial soups. He's more concerned with shorelines, the places where water meets land, creating intermediate surfaces which are neither bone dry nor soaking wet. Something rich and complex happens when air and water slowly erode tiny fragments of crustal rocks and wash them into streams and rivers, eventually to settle at the bottom of lakes and oceans as sediments. A typical river estuary is a transitional zone where erosion products haven't quite left the land. Yet neither have they been washed away to sink into the ocean depths. They accumulate into oozy, moist clay.

Clays have plenty of silicon, an element second only to carbon in the variety of compounds it can help to produce. These compounds usually contain aluminium and magnesium, sometimes iron, and always hydrogen and oxygen – but never carbon. When carbon-bearing organic compounds come into contact with clay, it can become a catalyst for further organic processes. The petrochemical industry has known about this for decades. As they drill down in their hunt for oil, they pump organic compounds that have the effect of making clays easier to penetrate.

Clay interacts in complicated ways with organic chemistry. One type of clay with an exotic-sounding name, **montmorillonite**, loosens in the

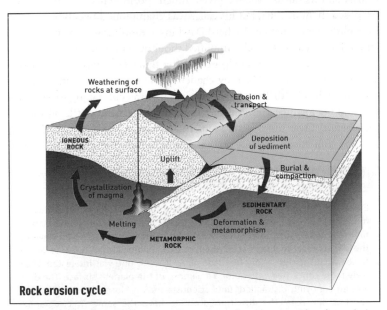

Rock erosion cycle

The supposedly solid rocky ground beneath our feet is reshaped and recycled over immense spans of time.

presence of certain organic compounds, yet locks solid in the presence of others. Its characteristics also change according to how much or how little organic material is pumped into it. Tannins, bitter-tasting brown compounds extracted from tea leaves and familiar for their use in the leather industry, are excellent clay looseners. (And talking of loosening, montmorillonite also works a treat for irritable bowel syndrome.)

Clay does something else quite remarkable. It reproduces. Clay crystals usually accumulate in porous flat sheets, like stacked roof tiles. When they dry out, the flat sheets easily flake away in the wind. Drop those flake spores into a moist environment and add more erosion deposits, and they create copies of themselves. They even transmit minor flaws and asymmetries in their molecular matrices from one generation of crystals to the next. Clay is fabulously complicated stuff. Although it isn't alive, it shares one or two crude similarities with life, and Cairns-Smith reckons it might be at least part of the solution to the chicken-and-egg problem of life's information content.

More clayful notions

Cairns-Smith's ideas weren't given much prominence when he first proposed them in 1965, but his cause was championed twenty years later by evolutionary geneticist Richard Dawkins. Combining their specialities, let's indulge in an extremely simplified thought experiment:

Suppose one clay type is so loose in water that its crystals easily separate and get washed away. Then, by chance, some crystals encounter simple organic molecules drifting in the water. The molecules cling to the crystals, and the fortunate result is that the crystals become a little bit stickier than before. Over time, water seeps in and out of the local clay deposit, and the non-sticky crystals are carried off while the organic-enriched

Martian clays

Earth is not the only planet in the solar system that makes clays. In the closing months of 1971, NASA's Mariner IX space probe arrived in orbit around Mars in time to witness a dust storm that threatened to block any interesting views of the planet below. Mariner's spectroscopes took readings from the atmosphere to see what the dust might be made of. At first, no one was interested. They were more concerned with getting a glimpse of the planet's surface. The dust was an annoying distraction, until scientists took a closer look at Mariner's spectroscope data. The dust turned out to have the same spectrographic profile as montmorillonite clay.

crystals stay put. A local density of clay–organics accumulates. Here we have the faintest dawn of selective advantage, the cold, hard driver of evolution turning simple entities into more complicated ones by virtue of what survives, and what doesn't.

Why don't we see clay–organic partnerships still at work today? In some ways, we do, as the oil companies know very well. Biochemists also continue to make fascinating discoveries about clay. In July 1996, James Ferris at the Rensselaer Polytechnic Institute in New York State persuaded nucleic acid-derived polymers (polymers are chain molecules made out of repeating elements) to condense around clay-like minerals, thus forming much longer chains than anything yet derived from traditional Miller–Urey soups. Ferris's article, impressively entitled "One-step, regioselective synthesis of up to 50-mers of RNA oligomers by montmorillonite catalysis", owed much to the example that Cairns-Smith had set more than thirty years earlier.

The main reason why natural organic chemistry no longer depends on clay is that DNA-dominated organisms have become faster and more efficient at sweeping up all available organic resources. Clays are too slow on the uptake, but perhaps our most cherished creation myths are not so far off the mark when they say we were fashioned from clay. Then again, clays might not have been especially important. Research in this area hasn't come up with any especially startling advances on Cairns-Smith's initial ideas. We don't know how life started on Earth. What we assume is that it didn't get going by divine magic.

The Goldilocks Zone

For those not familiar with the famous children's yarn, Goldilocks is a little girl who blunders into the cottage of three perfectly innocent bears and eats their breakfast. Daddy bear's porridge is too hot, mummy bear's porridge is too cold, but baby bear's is just right. When the bears return to their cottage and find that someone has been scoffing their food, they're curious as to the circumstances behind the theft. At the dawn of the space age, when the first serious searches for extraterrestrial life began, space scientists were similarly intrigued by Earth's good fortune in finding the right bowl of porridge, so to speak. It orbits the sun in a narrow belt of space, the **"Goldilocks Zone"**, that is neither too hot, nor too cold, but is apparently just right for liquid water – and therefore life – to exist. Mercury and Venus are too close to the sun, and unbearably hot. Mars is slightly too far away for comfort, and frigid cold.

In the 1960s, space scientists believed the Goldilocks Zone wasn't even as large as that special orbital realm. Even a planet as life-friendly as Earth would put constraints on where life could exist. Surely it couldn't thrive anywhere colder than the Antarctic, or hotter than the Gobi Desert, or darker than the uppermost levels of the ocean? Wouldn't life suffocate, burst apart in the low pressure, or be shattered by ultraviolet radiation if it dared to drift any higher than the rarefied atmosphere above mountaintops? Obviously life could not exist any deeper in the rocky Earth's crust than the realm delved into by plant roots and mud, or at least, the organic detritus carried into the lower depths by rain and meltwater?

All these assumptions are now being rewritten as we gain new and ever more startling knowledge of life in what would once have seemed exceptionally extreme environments. We have found microbes in nuclear reactors, microbes that love acid, and microbes that swim in boiling-hot water. Whole ecosystems have been discovered around deep sea vents which sunlight never reaches, and where the emerging vent water is hot enough to melt lead. We can no longer take it for granted that other life-bearing worlds must be even vaguely similar to our own.

Alien life on Earth?

Astrobiology is contributing new ideas to conventional biology, particularly the question of how hard – or easy – it is for life to begin. The assumption has always been that our existence, from primordial bugs to shopaholic sophisticates, is a staggeringly lucky fluke. What if life on Earth emerged not just once, but twice, or even several times? If this is the case, then the discovery of life on other worlds becomes even more likely. If we could find a terrestrial organism whose DNA and internal chemistry is markedly different from the kinds that we already know about, this would be almost as exciting as finding an extraterrestrial organism. Francis Crick often wondered about this. As he and Leslie E. Orgel wrote in a 1973 article, "It is a little surprising that organisms with somewhat different genetic codes do not coexist." Well, perhaps they do.

Mono Lake: life in a chemical soup

Mono Lake is located a few kilometres east of Yosemite National Park, near the town of Lee Vining, California. It has existed for at least 700,000 years, and was topped up by huge glaciers melting at the end of the last ice age. In 1941, the Los Angeles Department of Water and Power began diverting Mono Lake's water to meet the growing demands of Los Angeles.

Deprived of new freshwater sources, the volume of Mono Lake has halved in just a few decades, while its salinity has doubled. The levels of other chemicals in the water, such as arsenic, have also increased markedly. The arsenic is natural and not a product of pollution. What's changed is its concentration. As the clean water was drawn off by thirsty Los Angeles, the chemical silts and suspensions in the sludgy depths stayed just the same as before.

Mono Lake supports a simple yet productive food chain. At the bottom of the chain are microscopic single-celled algae, which are a food source for shrimp and small flies. They, in turn, feed nearly a hundred different species of water fowl. As the water's toxicity has increased, this ecosystem has come under threat. As far as the hunt for extraterrestrial biology is concerned, what we have here is a chemical soup that is becoming toxic to familiar, terrestrial forms of life. It should, therefore, make a good test-bed for what a hostile alien environment might be like.

GFAJ-1: a new microorganism?

What really excites astrobiologists is a species of bacterium in the water, a rod-shaped microorganism known as **GFAJ-1**. In December 2010, Felisa

Geobiologist Felisa Wolfe-Simon working at Mono Lake, where she discovered a unique microorganism that thrives in an arsenic-rich environment.

Wolfe-Simon, a NASA Astrobiology Research Fellow at the US Geological Survey in Menlo Park, California, claimed to have found the first example of a microorganism that apparently lives on a slightly different chemistry than all other life that we know about on Earth. She and her team published their Mono Lake findings in a paper called "A Bacterium That Can Grow by Using Arsenic Instead of Phosphorus" in the respected US journal *Science*. "Life is mostly composed of the elements carbon, hydrogen, nitrogen, oxygen, sulphur, and phosphorus", the team wrote. "Although these six elements make up nucleic acids, proteins, and lipids and thus the bulk of living matter, it is theoretically possible that some other elements could serve the same functions." Their startling claim was that GFAJ-1 "substitutes arsenic for phosphorus to sustain its growth".

Phosphorus is a component of the energy-carrying molecule adenosine triphosphate (ATP) found inside all biological cells. Phosphorus is also present in phospholipids, the waxy stuff that forms cell membranes, and it's a key element in the alternating sugar and phosphate "back bone" molecules of DNA. At the atomic level, arsenic is chemically similar to phosphorus, but it is poisonous for most life on Earth. "We know that some microbes can breathe arsenic", Wolfe-Simon said, "but what we've found is a microbe doing something new: building parts of itself out of arsenic." In the laboratory, Wolfe-Simon and her colleagues successfully grew microbes from the lake on a diet scarce in phosphorus and rich in arsenic. When phosphorus was removed entirely from the system, many of the microbes continued to grow and reproduce as if nothing had changed. One creature's poison (arsenic), is another's meat, so to speak.

Wolfe-Simon's results suggested that a bug can use arsenic in its very genes. Dozens of scientists assaulted her claims, arguing that GFAJ-1 may very well survive in arsenic-rich conditions by sequestering the element somewhere in its cell, but that there's no proof it actually incorporates arsenic into its structures, let alone its DNA. One commentator said that Wolfe-Simon had merely found arsenic clinging to the structures of GFAJ-1, just as mud clings to a boot without being part of the boot.

As always with microbiological specimens, preparing samples and analysing their exact molecular content is difficult work. GFAJ-1 is now undergoing comprehensive investigation as we try to settle the matter once and for all. If Wolfe-Simon's theory holds up, we would have proof, right here on Earth, that more than one kind of life is possible, even on the same planet. This would boost the probability of life on other worlds, perhaps living on chemistry even more wildly at variance with our own than GFAJ-1's seems to be. Carl Pilcher, director of the Astrobiology

What's the word for alien-hunting?

Nobel Prize-winning geneticist and microbiologist Joshua Lederberg (1925–2008) wasn't primarily a space scientist, but that didn't stop him from taking a keen interest in the possibility of alien life. In 1958, he helped the US National Academy of Sciences set up a Space Science Board, whose purpose was to map out a strategy for space exploration and human space flight, and to invent experiments that might search for life on other worlds. That was the year NASA opened its doors for business, as the US responded to Russia's launch of Sputnik in October 1957. Lederberg invented a new word that gave respectability to the hunt for alien organisms or their chemical precursors. He called the discipline "**exobiology**".

Another term for alien-hunting is "xenobiology" (biology of foreigners), though it's no longer used much. It was coined in 1954 by US science-fiction writer Robert Heinlein in his novel for younger readers, *The Star Beast*, the story of a high school student who discovers that his late father's alien pet is more than it appears to be. Xenobiology is no longer fashionable because of its associations with a similar word, "xenophobia", meaning distrust or hatred of foreigners.

In 1998, NASA adopted the term "astrobiology" to describe a broad interdisciplinary field involving geologists, astrophysicists, information theorists, molecular biologists and even philosophers. Indeed, it covers pretty much everyone with any serious interest in the search for life beyond the Earth. Astrobiology seems to be the winner.

Serious scientists also tend not to use the word "alien" in their discourse because it too easily conjures up images of science-fiction monsters. The preferred term is "extraterrestrial organism" or, in the case of an entity further up the evolutionary and technological scale, "extraterrestrial intelligence". In this book, we'll continue to use the word "alien", because we trust ourselves not to think of monsters, don't we?

Institute at NASA's Ames Research Center in Moffett Field, California, sprang to Wolfe-Simon's defence: "The idea of alternative biochemistries for life is common in science fiction. Now we know that such life exists in Mono Lake."

Looking for life on Earth – and failing to find it

It might sound crazy, but space scientists haven't yet found a reliable method for detecting life on Earth. If they can't find it here, they will have a tough job finding it anywhere else. When the Galileo space probe swooped past the Earth in 1990, building up gravitational momentum for its long trip to Jupiter, all its instruments were pointed towards us for a unique experiment designed to confirm the presence of life on our planet. Strong absorption of light at the red end of the visible spectrum, particularly over the continents, indicated the presence of chlorophyll, the molecule essential to (mainly green) plant life and photosynthesis. Spectral analysis of sunlight passing through Earth's atmosphere revealed its high oxygen content. Since oxygen is extremely reactive, a dead planet shouldn't support free oxygen in its atmosphere for very long. It has to be constantly replenished by life, especially from plants. Galileo also spotted small quantities of methane in the atmosphere.

However, in August 2003 a perplexed NASA team failed to sense life in northern Chile's Atacama Desert. Admittedly one of the world's driest deserts, it's teeming with life nevertheless. Scientists on the ground were pestered by flies, even as they marvelled at the variety of lichens growing on and under rocks, or watched vultures circling over their heads: a sure indication that plenty of other animals had to be around somewhere. But colleagues at the Ames Research Center in California, poring over photos and instrument data transmitted from the field scientists, never detected anything they considered unambiguous signs of life.

These results were unsettling for scientists trying to develop robotic landers and instruments. If an animal walked in front of a NASA rover trundling across the surface of another world, few observers would be left in doubt as to the presence of life, but what if it turns out to be sparse and harder to detect? What subtle data signatures, conveyed to us on fragile electromagnetic rumours from an unimaginably remote alien planet, might prove the existence of extraterrestrial biology? Today, at a time when new astronomical techniques offer clues about the atmospheres of several planets orbiting other stars, this question comes centre stage.

Is Earth alive?

Could our search for life on other worlds encompass worlds that are alive in their own right? In the late 1960s, NASA scientist James Lovelock was part of a team investigating the environment of Mars. As he pondered the differences between the inert Martian atmosphere and our own much more dynamic one, he realized that a fine balance exists on Earth as a result of constant feedback between living and non-living systems. He suggested that life maintains the oxygen composition of the air with unexpected precision. By 1979, he had broadened his ideas to include non-living rocks and ocean waters as facets of a planet apparently fine-tuning itself in favour of life.

Scientists now acknowledge that the Earth is indeed more than just an inert ball of rock with some life crawling around it. The fine balances may have been disrupted by excess carbon dioxide generated from human activity. Will Earth eventually shrug off her human tormentors, rather as an animal might ward off an infection?

Lovelock's **Gaia Hypothesis** is named in honour of an ancient Greek goddess, but he is the first to insist that none of the processes in his scheme are conscious or purposeful, nor mystical in origin. Think of how a mindless thermostat regulates the temperature in a room, multiply the complexity of that system a billionfold, and you have an idea of how Gaia functions. We are accustomed to metabolic stasis in biological organisms, be they ever so dumb. Self-regulation of temperature, acidity and energy levels is partly what distinguishes life from non-life.

We are also familiar with "symbiotic" plants, animals and bacteria that live in complex multidependent alliances, with each species relying on the others for survival. The Earth itself may be a super-symbiotic system. Words like "life" and "organism" are loose terms whose definitions are still hotly debated, and most scientists reject the idea that a planet can itself be alive, and doubt that the Earth is so stable as Lovelock supposes. Yet our home world does demonstrate some characteristics that would not surprise us if we observed them in a living creature.

The hunt for life on Mars

The twin Viking probes that landed on Mars in 1976 were the most advanced robots of their time. Their ingenious semi-automatic biological experiments sent back results that we still can't explain to this day.

Missions to Mars

Russian space planners could be forgiven for believing that Mars has a long-standing grudge against them. Despite beating the US into space by launching the first satellite and orbiting the first man, flying a probe to one of our nearest neighbours in the solar system has proved a more elusive challenge. Over the last decades they have attempted more than a dozen robotic missions to Mars, and none have gone smoothly. Six of them fell at the first hurdle, never even getting their payloads out of Earth's gravitational field. Two barely even left the ground.

Russia did earn the distinction of being the first nation to send an artificial object into the broad vicinity of Mars. In June 1963, Mars I (strictly speaking, the fourth probe after three failures) skimmed past the planet, as planned. Unfortunately it couldn't relay any of its findings, because radio contact had ceased three months earlier. Another probe skittered past Mars in February 1974 and missed its chance to fire its retro rocket and brake into orbit. All it could manage were a few hasty over-the-shoulder television pictures before disappearing forever into the depths of space... and so on and so forth, in a grim catalogue of failures for a country more used to space spectaculars than flopniks.

The Mariner IV space probe

NASA scientists had better luck, but in the early days of Mars exploration they might have wished the cameras had failed so they could hold on to their cherished dreams of Martian vegetation a little longer. On 15 July 1965, **Mariner IV** flew past Mars after a trouble-free journey of 230 days, coming to within 10,000km of the planet at its closest approach. In these earliest days of remote imaging technology, the transmission of pictures from deep space was no simple matter. Mariner's camera recorded 22 images onto a strip of photographic negative film, which was then processed internally on a convoluted series of rollers. A scanning device, similar in principle to a fax machine, then translated the results into radio signals for the waiting Earth, with pulses corresponding to light or dark areas on the negatives. Each picture took eight hours to transmit.

Mariner's radio gear beamed away for ten days solid to relay its slender batch of images. The result was our first close-up look at Mars in fuzzy frames of two hundred scan lines, each consisting of a string of two hundred tiny dots.

Nobody had seriously expected Mariner to encounter Percival Lowell's famous "canals", let alone the ruins of ancient Martian cities, but neither had the mission scientists expected to find craters – hundreds of them. Judging from these first views, it seemed that Mars might be, like the moon, pockmarked all over with impact scars, and just as inactive. At the very least, there had been some hope of finding gentle undulating plains, dune-swept deserts and signs of dynamic erosion processes.

The craters were a distinct disappointment. Their abundance, and the relative sharpness of their definition, appeared to indicate that nothing very much had happened on Mars since the craters had been formed many millions of years ago. Even more frustratingly, there were no obvious traces of vegetation. Patterns of dark areas in Mars's equatorial regions had often been observed from Earth through telescopes, and just as often attributed to vast tundras of lichen-like plant forms clinging to the beds of dried-out seas. Now they were revealed as nothing more than the most subtle differences in the shading of geological features.

Since the 1870s, successive generations of astronomers recorded apparently gradual changes in the shape of some of these dark areas, all of which had been attributed to seasonal spurts and fade-outs of plant growth. Mariner IV found nothing to support these intriguing ideas. Dust storms, rather than life forms, were responsible for the illusions of wide-scale surface change.

The Mariner programme confirmed that the Martian atmosphere consisted mainly of carbon dioxide, as most astronomers already suspected. But it turned out to be even thinner than anticipated, with barely one

Dramatic Mars

Mars has some of the most dramatic terrain in the entire solar system. The vents (caldera) of four colossal volcanoes are among the most obvious features. The greatest volcano, suitably christened Olympus Mons, covers an area the size of Arizona, is capped by a caldera that could swallow Hawaii, and climbs 27km into the Martian sky. Mount Everest, Earth's tallest peak, is less than half that height.

Mars's other dominant features don't climb, they plunge. Tourists flock to see Arizona's most impressive geologic scar, the Grand Canyon, a gorge 350km long and 24km wide in places, and nearly 2km deep. On Mars this would count as the merest scratch. Valles Marineris, a tangled network of canyons, gouges its way around nearly one quarter of the Martian circumference. The biggest trenches are more than 200km wide in places, and their steep walls extend 7km below the crust of the surrounding plains. These landscapes are on such a grand scale, they can only be appreciated properly from space. The slopes of Olympus Mons actually have a very gentle incline when viewed from ground level, while the walls of Marineris are so far apart that the opposing faces often cannot be seen at the same time. The geologic origins of these vast features are not fully understood.

The most colossal Martian feature is the least obvious. The major volcanoes, and large areas of rifts and fractures, are associated with the Tharsis Bulge. In this vast region, the planet's sphere has been pushed out of shape by the internal pressures of magma, molten rock beneath the crust that presses against the surface as if it's creating a giant blister. The four great Martian volcanoes were once fed by the blister's hot interior.

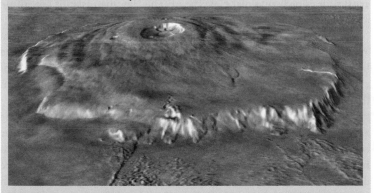

One of the most impressive features on Mars, the towering Olympus Mons volcano is the tallest known mountain in the solar system.

percent of Earth's atmospheric pressure. Any faint possibility that liquid water flowed on the planet was ruled out. In such a thin atmosphere, water molecules either freeze solid or boil away. When sunlight hits a patch of ice or frost on the ground, water molecules escape without becoming liquid first (a process called **sublimation**). The water clouds in the Martian sky consist of tiny ice particles, not steam.

Mariner IV's reconnaissance covered only a tiny proportion of the planet's surface. Believers in Martian life were not yet willing to abandon all their hopes. Such a small selection of fuzzy pictures could hardly be considered definitive. Who could tell what a broader mapping survey might reveal?

Earth and Mars compared

Earth has seven major tectonic plates (accounting for the seven continents and the Pacific Ocean) and many more minor ones. The mantle of molten rock immediately beneath them is constantly recirculating, causing the plates to drift very slowly, like rocky floes on an ocean of magma. This movement amounts to no more than a few centimetres per year, but over great spans of time measured in hundreds of millions of years, the Earth's landscape has been utterly transformed. Rifts appear between plates as they tear apart, and mountain ranges push upwards where plates collide. Plate tectonism moves active volcanic peaks away from underlying magma hotspots, the volcanoes eventually sealing up. Meanwhile the hotspots push up new volcanoes as fresh crust slides slowly into place overhead. The Hawaiian Islands constitute a long chain of volcanoes formed in this way.

In stark contrast, Olympus Mons and the other huge Martian volcanoes are quite isolated, so they must have remained in place above their magma hotspots throughout their periods of activity. The calderas' pattern of craters within craters betrays a long history of multiple eruptions from the same sites.

Whatever forces shaped the Martian landscape, they weren't the same as the plate tectonics that have gradually pushed apart our familiar continents, in the process reshaping not just the Earth's surface, but also much of the history of life upon it. Scientists are still arguing about the details, but we can be pretty sure that Mars's interior is cooler today than it once was. The biggest indicator is the weakness of its magnetic field in comparison to Earth's. As our world rotates on its axis, solid crust drags the iron-rich molten interior around with it. As a result, the gloopy interior's speed of rotation is slightly different from the crust's. A colossal dynamo effect produces powerful magnetic fields around the Earth. Luckily, these deflect nasty subatomic particles streaming towards us from space, especially from the sun.

Unfortunately, the sluggish Martian dynamo is jammed in the "off" position, so it lacks Earth's protective *Star Trek*-style radiation shield. This is not particularly good news for Martian life, but it doesn't necessarily have to be a game closer, especially for any organisms living under the topsoil, where they would be protected from excessive radiation.

The other big difference between Earth and Mars is sheer size. The Red Planet is actually quite a small world, half the diameter of Earth and with only a tenth of its mass. As a consequence, its gravitation is only a third as strong as Earth's. With

> **"Mars is not the Earth. As the legacy of Percival Lowell reminds us, we can be fooled."**
>
> Carl Sagan, *Cosmos* (1981)

so little attraction to hold it down, most of the Martian atmosphere has drained into space, slowly torn away by the relentless assault of solar radiation. The atmosphere today is whisper-thin, and the surface of Mars is bitterly cold.

Rivers on Mars?

Mars has no liquid water. This is odd because it does seem to have river-beds. On 1 July 1972, eight months into its mission, NASA's Mariner IX probe transmitted pictures of Mangala Valles, a 600km channel in the Southern Hemisphere. Unlike Valles Marineris, the jagged trench which cuts through a massive plain like a geological fault line, Mangala starts off as a fine network of tributaries covering hundreds of square kilometres of an upland area. These merge into wider, deeper channels that "flow" towards lower-lying territory. As the Mangala system reaches lower, and stretches further from the tributaries, it seems to thicken out. Some kind of fluid must have travelled down from higher ground, gathering mass and pace as it hit the lowlands. Other than its lack of water, Mangala is comparable in scale and characteristics to Earth's Amazon River. Intrigued, NASA scientists christened the downstream plain Amazonis Planitia in honour of this uncanny similarity.

The Mariner scientists began to think the unthinkable: at some stage in its 4.5-billion-year history, Mars might have supported colossal quantities of liquid water. And if there was water, then Mars must once have been much warmer than it is today, with a thicker atmosphere, otherwise the water would have boiled or frozen long before it could scar those great channels across the surface.

The Viking project (1975–76)

On the morning of 1 July 1976, the Smithsonian Institution in Washington, DC, opened the doors of its brand-new Air and Space Museum. A red, white and blue inaugural ribbon was stretched across the main entrance. President Gerald Ford stepped up in front of it, but he didn't make the crucial cut. A radio pulse from a tiny transmitter more than 100 million km away triggered an electric guillotine, and the ribbon was sliced by the very remotest of remote control.

Twelve days earlier, on 20 June, NASA's **Viking I** space probe had achieved Martian orbit after a flawless ten-month flight. The gimmick radio signal to Washington whetted public and political appetite for further treats to come. On 4 July, Independence Day (which in 1976 marked the two-hundredth anniversary of America's revolt against British rule), the Viking mother craft was supposed to send a lander right down to the surface of Mars, nailing national pride yet more firmly to the pages of history.

Keeping Mars clean

Planetary scientists are anxious not to confuse any biological data on other worlds with accidental infections from Earth. Robotic landers can be sterilized before launch, but any future human expedition to Mars would, inevitably, bring earthly bugs to the planet. Charles Cockell, a microbiologist for the British Antarctic Survey in Cambridge, and Gerda Horneck, an astrobiologist from the German Aerospace Centre in Cologne, Germany, are certainly not alone in suggesting that selected regions on Mars be preserved as the equivalent of "national parks", permanently protected against contamination or despoliation by robot soil-sampling arms or the rapacious digging of human colonists. In a 2004 edition of the academic journal *Space Policy*, the authors wrote, somewhat idealistically, "It is the right of every person to stand and stare across the beautiful barrenness and desolation of the Martian surface without having to endure the eyesore of pieces of crashed spacecraft scattered across the landscape."

Launched from Cape Canaveral on a Titan-Centaur rocket on 20 August 1975, Viking 1 embarks on its mission to discover whether there is life on Mars.

By the standards of 1970s technology, Viking was breathtakingly complicated. For the outward journey, the mother ship (the "orbiter") was attached to a smooth, white saucer-shaped canister, split into two distinct halves like a pair of inverted soup bowls joined together rim to rim. This formed a heat shield designed to withstand a fiery entry into the Martian atmosphere. Yet another layer of cladding, called the bioshield, protected the aeroshell and its carefully sterilized contents against contamination. NASA did not wish to send any terrestrial bacteria to Mars by mistake, in case microorganisms within the Viking hardware accidentally registered in the biological experiments and generated false signs of life on Mars.

Inside the aeroshell was the lander itself, like an insect larva in a cocoon, with its antennae folded into tight bundles and its three legs tucked tightly under its body. Such a comparison with a living creature is not entirely fanciful. This was a smart device with a rudimentary electronic intelligence. **Carl Sagan** of Cornell University, a planetary scientist renowned for his media popularizations of his field, worked at the heart of the Viking programme. He suggested that by some standards the lander was as clever as an insect, by others only as intelligent as a microbe.

"It takes millions of years to evolve a bacterium, and billions to make a grasshopper", he wrote in *Cosmos*. "With only a little experience, we were becoming fairly skillful at it."

This degree of intelligence was indispensable for Viking's safety. At the planned time of landing, radio commands would take just over nineteen minutes to reach Mars. Allowing for the equivalent response time from Earth, a 38-minute time lag had to be factored in. It was impossible for humans to guide the craft to a safe landing. Viking's computers needed to handle matters on their own.

The challenges of landing on Mars

The prime target site for Viking I was in Chryse Planitia, a region where the merging of four sinuous channels cutting through a plain seemed to suggest that water had once flowed there. Closer inspection of the favoured landing site from the orbiter cameras revealed nasty boulders. By definition, terrain flat enough for a safe touchdown is not particularly interesting. If an alien spaceship were to approach Earth for a landing, its extraterrestrial operators might gaze down wistfully at the artificial-seeming towers sprouting upwards in profusion from the great cities. The strange patterns of light, and the chaotic babble of radio stations and mobile phones, would arouse the dullest alien's interest. Alternatively, their probe might home in on the obvious potential of the Amazon rainforest, its green canopy visible from space, its humidity and balmy temperatures readily apparent. However, the best bet for a safe landing would be somewhere in the vast, arid, equatorial or arctic deserts, which appear fairly flat from an orbital vantage point.

Viking I faced these kinds of constraints. The most fascinating Martian terrains were almost always the most dangerous. The reality of accomplishing the landing wasn't quite like those pretty artworks that NASA had released in the preceding months. These had depicted the grey, three-legged billion-dollar lander nestling confidently among smooth undulating sands, with a couple of dozen harmless pebbles no larger than tennis balls scattered about at tasteful intervals. Sagan worked into the night, staring endlessly at the new pictures and wondering (as he wrote in *Cosmos*) if Viking I might be "condemned, like the legendary *Flying Dutchman*, to wander the skies of Mars forever, never to find safe haven." His colleagues wailed to a *National Geographic* journalist, "What if we land on a boulder? What if we sink into dust? What if we hit the side of a crater? This thing can't operate at more than a thirty-degree slope!"

Fortunately, a spare spacecraft acted as insurance against total disaster. Two ships had been launched atop a pair of unmanned Titan-Centaur rockets, three weeks apart, on 20 August and 9 September 1975. Because of its slightly different trajectory, Viking II was still just over a fortnight away from Mars on 20 July 1976, when NASA at last decided to try for a touchdown with Viking I on the safest terrain they could find within the Chryse Planitia target zone.

At 12.47am Pasadena, California, time on 20 July, seven years to the day since Neil Armstrong and Buzz Aldrin made the first moon landing on Apollo XI, flight controllers instructed Viking I's lander to separate from its mother ship and begin its descent. Three hours later, the vehicle reached the Martian surface intact, and six weeks later, the other set down in a similarly safe location, Utopia Planitia.

Carl Sagan: the popular face of space

Carl Sagan (1931–96) was director of planetary studies at Cornell University in Ithaca, New York, and played a significant role in NASA's deep space exploration programme from its inception. He was closely involved in the design and operation of scientific experiments on the Mariner II mission to Venus, the Viking Mars lander projects, and the Voyager, Pioneer and Galileo expeditions destined for the outer planets and their moons. He was among the first scientists to determine that the high surface temperatures on Venus are created by a runaway greenhouse effect, a phenomenon with sharp resonances for us on Earth.

This was the respectable academic guise of a man with a passion for telling ordinary people about the wonders of the universe via mass media. His 1980 television series *Cosmos*, and the accompanying book, introduced a worldwide audience to his characteristic phrase "billions and billions", a shorthand expression of wonderment for the sheer scale of the universe. He co-founded the Planetary Society, a 100,000-member international organization that campaigns on behalf of space exploration. Among his many books, *The Demon-Haunted World: Science as a Candle in the Dark* championed rational thinking, and argued against baseless superstitions and religious dogma.

Above all, Sagan was obsessed by the possibility of extraterrestrial life. He was instrumental in setting up the first formal radio searches for intelligent alien signals, and then protecting those early efforts against funding cuts. Sagan also wrote the thoughtful and plausible alien-encounter novel *Contact* (1985), which became the basis for a successful 1997 Hollywood film starring Jodie Foster. He died at the relatively early age of 62, from a rare form of leukaemia.

Viking's real work begins

The lander cameras at both sites revealed sandy landscapes peppered with countless small chunks of rock. The ground was rusty-red, just as everybody had anticipated. This was the Red Planet, after all. But the sky was a hundred times brighter than anyone had expected from such a thin atmosphere, suffused with a creamy orange-pink haze. You could mistake this effect for a cloudy, overcast day, except that Mars can't support dense clouds. In fact, a permanent suspension of very fine dust particles reflects the sunlight, giving the Martian atmosphere its gentle glow.

Now came the really hard part. Viking's primary mission was to try to answer one of our greatest questions. Is there life on Mars?

NASA designed a suite of experiments that could ask this question automatically, in four different ways, while operating 100 million km away from the nearest scientist. Somehow, the equivalent of a biochemistry department in a well-funded university had to be packed into a space less than the size of a car battery, and made to work without technicians and faculty chiefs. Within this compact box, carbon was the star of the show.

Taken on 20 July 1976 by the Viking lander, this was the first image ever returned from the surface of Mars.

Finding life: the starting point

Despite the cliché that rocket science is hard, at least the problem of landing a probe on another world is definable in a way that brooks little argument. It's just a matter of breaking down the overall task into discrete sub-tasks, phrased in terms of specific sensor readings. What's the height above the ground? X, as determined by radar. What's the rate of descent? Y, as calculated from shifts over time in the radar signals pinging back from the surface of the onrushing target planet. What's the necessary braking thrust? Z, as determined by readings of the rocket engines' temperatures and pressures. When do you shut off the landing engine? (The moment when X = zero is usually considered to be a good time.) X, Y and Z can be quantified as values in a fairly simple algebraic equation loaded into the guidance computer.

Rocket voyages can be modelled in digital form, ahead of a mission, and encoded within a planetary lander's guidance systems. But that doesn't work for a device whose task is to search for life. Astrobiologists can only dream of having such simple problems to solve. What specific instruments, and what kind of data delivered from them, might produce suitable X, Y and Z values for the detection of alien life? It is impossible, for instance, to be sure that life on Mars, or any other extraterrestrial world for that matter, will share any characteristics with life on Earth. If it doesn't, tests may become meaningless, since the results can't be compared against any sensible yardsticks, with no X, Y and Z of known values to serve as our guide.

Where to begin in the search for life on Mars? A hunt for animals, plants, or any organisms with an obviously visible structure seems unrealistic, now that all our 1950s science-fiction dreams of bug-eyed aliens with antennae on their heads have faded. A lander's camera system cannot be expected simply to capture convenient snapshots of life forms crawling around it. No one today seriously anticipates finding anything larger or more complex on Mars than single-celled organisms. At this tiny scale, it was better to focus on the kind of measurable chemical activity that Martian microbes might demonstrate, thus betraying their presence indirectly.

When it comes to predicting what tell-tale traces to look for, some sweeping guesswork is required. Just look at the problems of analysis we face closer to home. There is a wide variety of terrestrial microorganisms whose chemistries use neither plant photosynthesis nor animal oxygen respiration. Indeed, there is a whole class of anaerobic bacteria which don't just fail to thrive, but swiftly become depressed, then downright dead, in the presence of the oxygen that keeps the rest of the biological world alive.

Viruses are a further complication. Like all living things they contain nucleic acids which govern their replication. They multiply in the right conditions (as parasites within the cells of other organisms), but don't respire, eat or excrete, meaning they emit no exhaust products that might betray their presence to eager instruments. In fact, nobody is quite sure whether or not to regard viruses as living entities. What if a probe finds something on another world with virus-like qualities, neither definitely alive nor obviously inert? Could any instrument detect such entities if they don't give off chemical traces? To round off the list of problems, we might encounter forms of life totally beyond human recognition. Something, quite literally, alien to us.

Carbon chauvinism

But the quest has to start somewhere. Astrobiologists tend to focus on organic (carbon-based) chemistry, the one sure-fire common denominator of all life on Earth. All living things absorb, manufacture, reconfigure and expel molecules based around carbon. Some of them, such as carbon dioxide or methane, are extremely simple. Others, such as DNA, are dauntingly complex. Even so, nothing happens within a living entity unless carbon is in the molecular mix. Even viruses are organic in their construction and replication.

Sagan admitted in 1975 that "it's hard not to be a carbon chauvinist". That same year, his colleague Gerald Soffen, one of the lead scientists in NASA's Viking Mars lander project, explained to the *National Geographic*, "Carbon is incredibly flexible. The atoms can make long chains, and they can attach to other atoms in an endless number of configurations. Only carbon can provide the incredible variety of molecules needed by any living organisms we can conceive of." Today, most biologists still share this intuitive belief that organic chemistry is probably fundamental to life across the universe as a whole.

What's so special about carbon?

Carbon atoms have six electrons divided between two electron shells. The inner shell holds two, while the outer valence shell contains four. A carbon atom is predisposed to bond with as many as four other atoms. This makes it possible for long molecular rings and chains to be formed with other atoms, such as hydrogen, oxygen, nitrogen and sulphur. More than sixteen million compounds of carbon are possible, a far greater number than can be centred around other elements. Silicon, an element easily available in the Earth's crust and an essential component of clay minerals, is capable of forming similarly diverse bonds and molecular structures, but silicon-based chemistry is hindered by needing higher energies – more heat, essentially – to trigger it. Carbon can perform extraordinary feats of assembly and synthesis at relatively low temperatures.

Almost everything we use as fuel, whether in the form of food for us, coal for the fire or petrol for cars, is based on one kind of carbon-based molecular chain or another. These molecules, in turn, are derived from living chemistry, either current or active in the distant past. In the non-living world, carbon's value as an engineering material is due to the fact that it can bond in many different ways just with other carbon atoms. For example:

▶ **Graphite**, the "lead" in pencils, is a useful lubricant for machinery because its carbon atoms are loosely bonded in two-dimensional crystalline layers which easily slide over one another.

▶ **Carbon atoms in a diamond**, by contrast, are arranged with strong bonds in all three spatial dimensions, making it the toughest naturally occurring substance.

▶ **Crude oil**, the viscous black-brown stuff that emerges from an oil drilling field, is a complicated substance containing a mixture of thousands of different hydrocarbons, molecules consisting of nothing but hydrogen and carbon atoms.

Key carbon compounds

Carbon compounds can also verge on the biological without actually constituting anything that can be thought of as alive. Scans of deep space reveal that interstellar nebulae, the clouds of dust and gas drifting between the stars (and themselves the product of previous stars), contain hundreds of different chemicals, including amino acids and simple sugars manufactured by ultraviolet radiation from stars acting on molecules within the clouds.

▶ **Sugars** are among numerous types of carbohydrates. Not to be confused with hydrocarbons made purely from hydrogen and carbon, carbohydrates contain one additional element: oxygen.

▶ **Amino acids** are a little more complex. All aminos contain carbon, oxygen, hydrogen and nitrogen, and some also incorporate sulphur. Amino acids always consist of a basic amino group (NH_2) and an acidic carboxyl group (COOH), with a third "R" group that can vary widely in molecular structure. Carbon atoms are the essential links between all the groups. Upwards of four hundred different amino acids have been identified, of which twenty in particular make up the assembly units for all proteins, the first level of organic molecules that are directly associated with living things, and which cannot exist independently of life.

▶ **Proteins** are very long chains of different amino acids, the ordering of the aminos determining the shape, identity and functioning of each protein. The human body contains at least two million different proteins. Animals and plants between them make use of at least ten million types of protein. This is where the simplicities of non-living organic chemistry are left far behind, and the complexities of life begin. When we find amino acids in a meteorite, we get mildly excited at this hint of a pre-biological building block, but if we ever encounter a protein that's not from Earth, then the "life elsewhere" deal is done.

Where carbon's found

We find carbon wherever we find materials or deposits that have had some interaction with the natural organic world. Even something as apparently artificial as plastic is made from hydrocarbons obtained from fossil fuels. Carbon moves restlessly between the land, atmosphere and oceans, and also migrates in and out of the Earth's crust, swallowed by tectonic shifts and regurgitated by volcanism and erosion, or by human activities, such as drilling for oil and gas.

The biological phase of the carbon cycle takes place every moment animals and plants, living or dead, absorb or release carbon into the surrounding environment. In the atmosphere, carbon is attached to oxygen in carbon dioxide (CO_2). Plants absorb it and split the atoms, holding on to the carbon and expelling the oxygen. Animals eat the plants or eat the plant-eaters, but whatever their chosen diet, they end up absorbing the carbon. Every time an oxygen-breathing creature such as ourselves breathes out, a puff of carbon dioxide gas is released, putting carbon back into the atmosphere. Many animals and plants also expel methane (CH_4), another carbon-bearing gas.

When plants and animals die, their tissues decay and their carbon ends up in the ground. If successive layers of decay products build up over long time spans, and are subsequently buried by additional geologic processes, the carbon can be trapped underground for millions of years as fossil fuel deposits. Carbon also accumulates on oceanic seabeds as the shells and bones of marine animals and plankton collect on the sea floor. The carbon in these dead husks accumulates in chalk and limestone deposits. Whether in the living or the dead, carbon seems to be an inevitable component of biology. Little wonder that the Viking scientists had it in mind when they designed their biology experiments.

The Viking space probes' hunt for carbon

The first search for life beyond Earth was initiated at the Chryse Planitia landing site on 28 July 1976, when Viking I extended its robot arm and scooped up a fistful of Martian soil. Because of the significant radio time lag between the transmission of commands from Earth and their reception by Viking's antenna, "real time" human control of the experiments was impossible. Mission controllers had to depend on Viking to

A key member of the Viking design team, planetary scientist Carl Sagan stands proudly by a model of the lander in Death Valley, California.

haul in the soil automatically, and distribute portions evenly between the various compartments of the laboratory box. Once the samples were in place, all the experiments (outlined below) were activated simultaneously, because they had to share key pieces of the general hardware, such as the heaters, gas pumps and data tape recorders. Essentially, Mars was "asked" four questions:

Does anything in your soil exchange gases with your atmosphere?

The Gas Exchange (GEX) experiment was designed by Vance Oyama and his colleagues from NASA's Ames Research Center in California. It detected any transfer of gases, whether organic or inorganic, between the Martian soil and the surrounding atmosphere. Very gently, the soil was exposed to a humid – but not soaking – broth of amino acids, vitamins, some potentially tempting carbohydrates and a few inorganic salts. The assumption was that if any bugs were alive on the planet's surface, they would have adapted long ago to an arid lifestyle. A sudden influx of moisture might shock them to destruction. Then the atmosphere in the tiny test chamber was sampled at regular intervals to see if anything in the soil had processed the nutrients and expelled waste products.

GEX used a gas chromatograph, a column containing a filtration unit that sifted molecules according to size. Certain types of molecules would pass through very fast, others more slowly, just as a drop of ink seeps across a sheet of blotting paper and separates out into its different constituents. The hardware could distinguish simple chemicals like hydrogen, oxygen, carbon dioxide, nitrogen and methane, all of which can be found in Earth's microbial waste products.

The first GEX results from Chryse Planitia electrified the Viking scientists back on Earth because the soil sample gave off a tremendous burst of oxygen, boosting the overall pressure in the test chamber by a factor of five. "Never have we seen that magnitude of response in any of the Earth samples we tested that did not contain life", said Viking scientist Dr Gilbert Levin at the time. But the oxygen burst tailed off, and further tests with the GEX didn't create any more jolts of excitement.

Does anything in your soil release carbon?

The Labelled Release (LR) experiment was designed by **Dr Gilbert Levin**. A soil sample was placed in a container, along with a volume of uncontaminated

Dr Gilbert Levin: the Viking rebel

Dr Gilbert Levin was unusual among NASA space scientists because his background was in commercial health rather than space science. After serving in the US Navy towards the end of World War II, he took a master's degree in sanitary engineering, and served as a public health engineer in Maryland and California before starting an environmental consulting firm, Resources Research, in 1955. Three years later, he happened to meet NASA's chief administrator at a cocktail party, and mentioned his lifelong fascination with Mars, as well as an idea for detecting any microbial life there.

NASA contracted him to refine that idea, and in 1969 it was formally selected for inclusion in the Viking project as the Labelled Release (LR) experiment, a clever miniaturized adaptation of Levin's patented method for detecting potentially dangerous bacterial contamination in water supplies, foodstuffs and hospital blood banks, even in very low concentrations. Levin's Viking team also included Dr Patricia Ann Straat, a young microbiologist from Johns Hopkins University (see p.69).

Now in his late seventies, Levin is still active in business, and still insists that "the Labelled Release life detection experiment aboard NASA's 1976 Viking Mission reported results which met the established criteria for the detection of living microorganisms in the soil of Mars".

Martian atmosphere. Then the soil was humidified with a fine mist of nutrients laced with radioactive carbon-14 from Viking's special reserves. Chemically, this stuff would behave the same way as any other carbon, except that Viking's instruments could keep tabs on it, whether going into a prospective bug via the nutrients, or coming out in waste products. If carbon-14 went into the soil sample and never re-emerged, that would be a dull result. On the other hand, if anything in the soil puffed out radioactive whiffs of waste, that would be the opposite of dull.

Levin argues to this day that his LR experiment yielded the most positive of all Viking's biology results – so positive, indeed, that they were consistent with finding life. As soon as nutrients were introduced, the Geiger counter registered atom after atom of radioactive carbon given off by the soil, producing 9000 blips per minute for 7 Martian days (each of which lasts nearly 40 minutes longer than our own 24 hours). Separate control samples of soil were heated to destroy any prospective organic compounds before the LR tests were conducted on them. After heating, the unusual LR reactions were not in evidence. Something in the soil seemed to show the same kind of sensitivity to heat that we might expect of organisms.

Does anything in your soil take in carbon?

The Pyrolytic Release (PR) experiment, devised by Norman Horowitz from the California Institute of Technology, was the inverse of Levin's LR experiment. It looked for carbon going in, not coming out. It was aimed at finding life forms that might breathe in carbon dioxide gas from the Martian atmosphere and incorporate the carbon into their metabolism.

The PR's test chamber was illuminated on the inside, in simulation of sunlight. No nutrients were added to the experiment, since the focus was on plant-like chemistry, under the assumption that Martian microbes might take all they needed from the soil, carbon dioxide atmosphere and sunlight without any need for extra food.

Carbon dioxide from Viking's personal supply, artificially laced with radioactive carbon-14 tracer atoms, was injected into the chamber, entirely replacing the natural Martian gas. The PR system ran for five days, after which all the gas in the top of the chamber was pumped out, so that any remaining loose carbon-14 would be safely removed from the equation. Then the soil sample was heated to just over 600°C, with the simple and merciless aim of tearing apart any delicate organic structures and sending their atomic smoke towards a radiation counter. If any carbon-14 emerged from the wreckage, this would suggest that something in the soil had absorbed it during that five-day run. The results suggested a very low level of biological activity. The amount of "fixed" radioactive carbon detected in the soil was marginal, but it was there.

A fresh control sample of soil was exposed to the PR system, this time heated to a point beyond life's comfort level before tests began. At the end of this run, the soil's absorption of carbon was a fraction of the earlier sample's. Apparently something was badly damaged, if not entirely eradicated, by the heating. "You could have knocked me down with one of those Martian pebbles", said Horowitz at the time.

Although the initial PR results were encouraging, Horowitz didn't want to commit himself. "I want to emphasize that we have not found life on Mars", he told journalists at a NASA press conference. "The data we've obtained is conceivably of a biological origin, but biology is only one of a number of alternative explanations that have to be excluded. We hope to have excluded all but one of the explanations, whichever that may be."

Donald Rumsfeld with his famous "known knowns and known unknowns" might have understood Horowitz's statement, but journalists at the time weren't sure if that was a "yes" or a "no" as to the question of life on Mars.

Does your soil contain any organic compounds?

Viking's most important test was the one that rained on everyone's parade. The Gas Chromatograph/Mass Spectrometer (GC/MS), supervised by Klaus Biemann from the Massachusetts Institute of Technology, was supposedly the most critical of the experiments. Its hardware was complicated, with a canister completely distinct from the GEX, LR and PR assemblies, its own privileged supply of soil, and a device for grinding the sample down into tiny fragments that could easily be vaporized.

A soil sample was heated to release any organic substances into it as vapour. These products were pumped past a small beam of charged particles, fired from the same kind of device you'd find in the back of an old-fashioned TV tube. This bombardment knocked off some electrons from the molecules as they passed through, leaving them with a slight positive charge because of the protons that remained in their atomic nuclei. Now they could be deflected by magnetic fields. Heavier molecules were less bothered by the magnets than lighter ones. At the end of their dramatic ride, each molecule slammed into a screen of electronic detectors, hitting different positions on the screen according to their mass. With this quite staggering piece of 1970s space technology, Viking would supposedly be able to sniff out organic compounds. It found none. Viking had failed to deliver its most important clue.

Not taking "no" for an answer

Long after the Viking programme came to a close, Levin and his colleague Patricia Straat continued to claim that the mass spectrometer hadn't been sensitive enough to find organic compounds on Mars. While NASA as a whole presented a common front, announcing with regret that it hadn't found any compelling evidence for Martian life, Levin departed from the official party line, and claimed that life had in fact been detected. He began a three-year hunt for proof that the mass spectrometer wasn't good enough.

Before Viking's experiments were loaded on board the spacecraft for their long flight to Mars they had, of course, been tested on Earth. A variety of samples was analysed, from warm sods of Californian lawn to

"To this day, no non-biological replication of the complete Viking data has been achieved."

Dr Gilbert Levin, *The Microbes of Mars* (2011)

frozen Antarctic tundra. Exposed to these benchmark samples, the GEX, LR and PR tests all demonstrated fine sensitivities, but the relative scarcity of microbes in some of the frigid Antarctic materials proved a tough challenge for the mass spectrometer. One sample in particular failed to register in this experiment, yet bleeped happily in Levin's radioactive LR test. Levin didn't think it was fair to dismiss his startling LR results just because the mass spectrometer had failed to find organic compounds in the cold, dry, hostile environment of Mars. Viking's apparently negative overall results proved nothing, Levin claimed.

Complicating the questions

When NASA was coming up with biology experiments for the Viking programme in the early 1970s, microbiologist **Wolf Vishniac** (1922–73) thought it was dangerous to make any specific chemical assumptions about Martian life. He devised an experiment that made no detailed analyses whatsoever.

For his "Wolf Trap" experiment, a small sample of Martian soil would be dropped into a transparent jar, and a translucent nutrient broth added. Just one type of measurement would be taken before and after the broth was added, and then at regular intervals over time, to monitor how translucent the broth remained. A calibrated beam of light passing from one side of the jar to the other was all that was required. Vishniac argued that if the fluid in the chamber clouded up at any stage, then some kind of microbial growth within the broth could be assumed. He did not think it necessary, or even desirable, for Viking to investigate the exact chemical details.

In November 1973, Vishniac set out to test his trap in a place that approximated, as closely as our planet can manage, the harsh conditions on Mars: the Antarctic. Vishniac selected an intensely cold and arid valley in the Asgard Range, close to Mount Balder, in the west of the ice-bound continent. Here he placed a selection of his traps into the ground. He had mixed a "low-calorie" nutrient suited, he believed, to such a tough environment. On 10 December 1973, Vishniac set out to retrieve his samples. Sadly, this was the last time he was seen alive. At some point on his solitary walk he must have lost his footing. Later that day, a search party from the McMurdo research station found his body at the base of an ice cliff.

Vishniac's grief-stricken colleagues determined to retrieve his samples on his behalf. The results were astounding. His nutrients were fogged with microbes from an apparently sterile Antarctic region previously considered unsuited to life. Vishniac's widow, Helen Simpson, identified a new species of yeast unique to the Mount Balder area. Other than scaling down the richness of the nutrient in keeping with the scarcity of food in the target site, Vishniac had made no assumptions about Antarctic biology. He had simply designed an experiment to find some. If Helen Simpson had not been on hand to identify the yeast, a suitable light beam sensor could have registered the clouding in the solution and a robot probe could have beamed back a message to a remote base. That message would have read: "Life is here."

NASA's managers tend to design complicated projects. There's some hard-headed political reasoning in this. Simple experiments and cheap spaceships don't involve a sufficient number of researchers and manufacturers. Consequently it can be hard to obtain the backing for small-scale missions. By devising expensive and sophisticated procedures and hardware, NASA ensures that many different organizations can profit from space activities, either scientifically or in terms of construction contracts. In this way, a moderately wide spectrum of political support for funding can be achieved. Such logic might partially explain why Vishniac's charming, simple experiment was dropped from the Viking programme as a result of budgeting problems. The Wolf Trap was too cheap, rather than too expensive. But there might also have been an image problem. It would have been difficult for Viking's political managers to justify sending a test to Mars whose result, at best, would have been: "Yes, there's some kind of life. The nutrients went a bit cloudy." The scientific community would want much more rigorous data to fuel their pet theories.

There's no doubt that the Viking team *did* get plenty of data. The twin landers were equipped to undertake a wide and detailed variety of measurements, and a wealth of information was gathered, but the very breadth of analytic capability proved deeply confusing when the results were transmitted to Earth. After all that effort, we still don't know whether or not there's life on Mars.

Superficial answers

The problem with the Viking space probes' surface investigations is that they were, quite literally, superficial. Mars is covered with fine dust and sand. The dust gets everywhere, including into the smallest cracks and crevasses in boulders. Mars's reddish hue is mainly a result of that dust, created from rusty iron-rich minerals that presumably formed long ago when Mars was warm and wet. Now that the planet is cold and dry, ultraviolet solar radiation has played strange tricks on the dust, creating superoxides. Oxygen atoms in the "rust" are accompanied by loose electrons urgently seeking to reconnect with any atom that happens to have a vacant slot in its electron configuration. This extremely unstable set-up makes Martian dust sensitive to any change in the local environment, such as the arrival of moist organic nutrients from a spacecraft. The strange bursts of activity that so excited Viking scientists were probably created by this extreme reactivity, not Martian bacteria.

But there's more to Martian soil than toxic superoxides alone, and more variation in the planet's surface characteristics than the two Viking sites suggested. The reactive dust is just a thin layer on an otherwise much richer planet. When NASA's Phoenix lander arrived in May 2008, touching down in the northern polar region of Vastitas Borealis, it confirmed the presence of water ice. It also found useful nutrients in the soil, such as magnesium, sodium and potassium, all of which are necessary for living organisms.

Initial reports from Mars dampened hopes of finding environments hospitable to life. Carl Sagan, usually the most ebullient of the Viking scientists, was curiously subdued when the first pictures came back from the landing sites. It was obviously a disappointment that the two regions, separated by a whole planet's width, one of them broadly equatorial and the other in high northern latitudes, had turned out to be so similar. Suppose a pair of Viking robots had touched down on Earth at equivalently separated places, say, a Nebraska cornfield and a patch of Icelandic tundra. There would be many shared chemical characteristics, but some noticeable differences would have emerged. For instance, Nebraska farmland would have proved much richer than Icelandic tundra in nitrogen and phosphates. No such obvious contrasts were revealed by the two Vikings. Sagan's mood improved when it dawned on him that "four instruments on the landers were designed to detect life, and three of them produced positive signals. What does this mean? Even if it turns out that the results are non-biological, then there are obviously chemicals that simulate the behaviour of life just lying around in the soil."

Even if there is no life on Mars, it's still worth finding out more about the Red Planet. Earth and Mars were both shaped from the same solar system stock, and there's no particular reason why Mars should have lost out in the great game of life, other than losing its atmosphere to space because of its comparatively smaller size and weaker gravitational field. Mars might have held out long enough for something interesting to happen, even if life never actually emerged there. Further exploration of its soils, rocks and dried-up riverbeds might reveal what Earth was like at a time when plants and animals had not yet scattered their kind throughout the oceans and across virtually every scrap of land.

Roving on Mars (1997–)

Today's technology makes Mars seem almost as close and accessible as our own backyards. High-definition images of its surface can be downloaded on home computers within a few days or even hours of transmission. Ground controllers need time to process the images, but essentially they are made available to the public as soon as they are ready. We take for granted our god's-eye access to Mars, Jupiter, Saturn, dozens of moons and even several major asteroids.

The Mars **Pathfinder** mission, NASA's first touchdown project since the Viking era, opened a new era of Internet space exploration. As Pathfinder

Looking for signs of life: the Curiosity rover will be capable of detecting the slightest trace of organic chemicals on the surface of Mars (artist's concept).

Timeline: the highs and lows of Martian exploration

1962: Russia's Mars 1 is the first probe to pass Mars, but contact is lost.

1964: NASA's Mariner 4 is the first successful Mars mission, returning 22 TV pictures of the surface.

1969: Entering different orbits around Mars, the twin Mariner 6 and 7 spacecraft take photos of equatorial and polar regions, and confirm that Mars's atmosphere is comprised mainly of carbon dioxide.

1971: Russia scores another brief victory when its Mars 2 craft releases the first human-made object to reach the surface. But the lander module crashes, and the orbiting mother ship's view of the surface is blocked by a dust storm.

1976: NASA's twin Viking landers and orbiting mother craft do rather better, mapping the planet in detail from above and searching, inconclusively, for life on the surface.

1993: NASA's Mars Observer spacecraft is lost during its final orbital approach, possibly because of a small onboard explosion.

1997: NASA's Mars Global Surveyor successfully uses aerobraking in the upper regions of the atmosphere to achieve the low orbit required for its detailed mapping mission from space, which lasts for nine years before contact is lost. That same year, NASA's headline-grabbing Mars Pathfinder becomes the first touchdown craft to be launched in two decades. Renamed the Carl Sagan Memorial Station soon after its successful arrival, the lander releases a small wheeled rover called Sojourner.

1999: Japan's *Nozomi* (Hope) probe misses its target, ending up in a long, lonely orbit around the sun. NASA's Mars Climate Orbiter disappears behind the planet and is never heard from again, possibly due to navigational errors causing it to fly too close to the atmosphere. And in yet another costly foul-up, the Mars Polar Lander switches off its braking engines too soon and crashes.

2001: NASA's Mars Odyssey makes it safely to Mars and begins mapping the surface distribution of minerals and water ice. The probe is still operational.

2003: Britain's Beagle-2 lander is lost, probably as a result of a crash, although its much more capable parent craft, the European Space Agency's (ESA) Mars Express, functions flawlessly in orbit.

2004: NASA's Mars Exploration rovers, Spirit and Opportunity, arrive on Mars and begin their long and brilliantly successful surface missions.

2006: NASA's Mars Reconnaissance Orbiter successfully exploits the friction of Mars's upper atmosphere to slow down and achieve orbit, at the start of a high-resolution survey mission.

2008: NASA's Phoenix spacecraft finds water ice and potentially nutritious minerals at its landing site on Vastitas Borealis, a lowland region near the Martian North Pole.

and its little wheeled rover Sojourner beamed back startlingly clear images of the planet's surface when it landed on 4 July 1997, NASA's websites absorbed over thirty million hits. Larger and more advanced rovers, Spirit and Opportunity, landed safely in January 2004 and operated with extraordinary success for over half a decade. They identified rocks and soils that confirm plenty of liquid water activity on the Martian surface at some time in the past, but weren't equipped with biological experiments.

A fourth and even larger wheeled rover, the Mars Science Laboratory rover Curiosity, is expected to land in August 2012. It will carry a more advanced version of the gas chromatograph used for the Viking project, and will be capable of detecting the slightest trace of organic chemicals in the soil or the surrounding atmosphere. The European Space Agency (ESA) is building a similarly ambitious rover, due for launch in 2016 as part of its ExoMars programme.

A veritable fleet of orbiting probes has surveyed Mars from above, mapping it in excruciating detail and confirming the presence of water ice, salt beds stranded when ancient bodies of briny water evaporated, and large, thick clay deposits, another key indicator of an aquatic past. The Martian sky is also coming up with surprises. In December 2003, the European-built Mars Express orbiter detected significant traces of methane in the planet's thin veil of atmosphere. More than nine-tenths of terrestrial methane (a hydrocarbon consisting of four hydrogen atoms bound to one of carbon) is a by-product of life, whether in the form of fossil fuels and rotting swamps, or puffing out from the backsides of cows. The small fraction not produced biologically is geologic.

Intriguingly, the Martian methane is unevenly distributed, and its concentrations change with the seasons. This gas can't be a holdover from ancient times when volcanism was rife on the Red Planet. Ultraviolet radiation from the sun quickly breaks down methane, whether on Earth or Mars. Its continuing existence suggests sources that are still active to this day. In theory, it could be a waste product from microorganisms living under the Martian ice or buried deep under the soil.

Multiple methane plumes have been observed, one of which released nearly 20,000 tons of the gas. The plumes seem to peak during the warmer seasons, possibly as a consequence of permafrost easing up a little and allowing methane to seep into the atmosphere, but this is pure speculation. Among its many tasks, the Mars Science Laboratory will measure isotope ratios in the methane, looking for that all-important dominance of lightweight carbon over its heavier cousin.

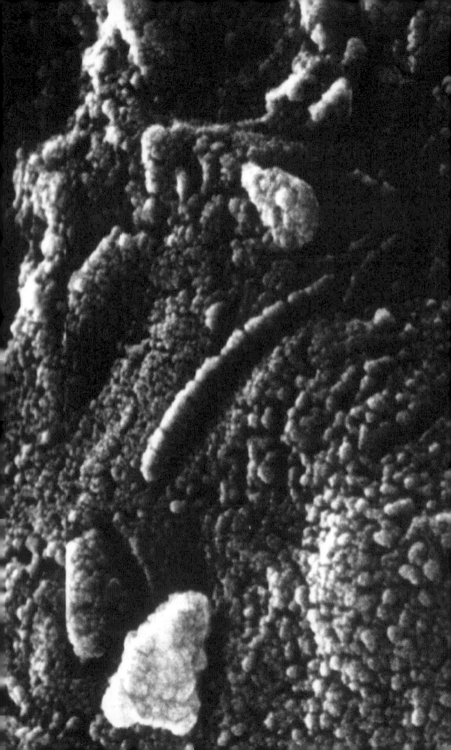

Mars reawakens

The Viking biology experiments of the mid-1970s were disappointing after decades of anticipation about life on the Red Planet. All bets were off again two decades later when a small chunk of Mars was discovered on Earth, apparently laden with the fossilized traces of ancient Martian life.

Rogue rocks

More worlds than we can see today condensed out of the solar system's accretion disc. It is likely that there were several planetesimals: objects much larger than mere rocks, yet too small to be counted as planets. At least a few of them must have developed orbits around the sun that were too elliptical (egg-shaped) for long-term survival. At some point in the distant past, their paths intersected and they smashed together, creating a swarm of rubble that now drifts in a chaotic 550 million km-wide band of space between the orbits of Mars and Jupiter; it's known as the **asteroid belt**.

This belt may even represent the remains of a quite sizeable planetesimal, torn apart by the tidal forces exerted by Jupiter's vast gravitational field. So far, we're not absolutely certain where the asteroids came from, but we can be pretty sure that they are the smashed remnants of larger objects.

Scientists tend to categorize asteroids into two main groups, based on how they appear in infrared telescope images. The brightest-looking asteroids (brightest in infrared, that is) are rocky metal-rich bodies, with high concentrations of iron and nickel. The darkest asteroids contain high quantities of carbon – and some of these are also suspected of harbouring water ice. Nearly three hundred asteroids have been classified

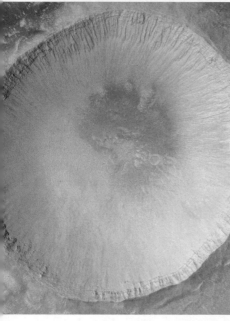

Visible asteroid impacts in Arizona (above, 1.1km wide) and on Mars (left, 3.6km wide) are small fry compared to the massive crater hidden under the Yucatán Peninsula (200km wide).

as potentially hazardous by the astronomers who track them. One day, some of these might be pulled away from the asteroid belt by gravitational influences within the solar system. If so, they could end up on a collision course with us. The Earth has been hit by rogue asteroids many times in the past, and it's bound to happen again in the future.

The big one

Approximately 65 million years ago, an asteroid the size of a mountain struck the Earth near the coast of Mexico. The impact caused a huge explosion, gouging out a crater more than 200km wide centred on the Yucatán Peninsula. The shock waves alone, in both the atmosphere and the oceans, must have caused catastrophic damage across vast swathes of the Earth's surface. Dust hurled high into the atmosphere obscured the sun for many months, and possibly even years, before dissipating. Light and heat became scarce for a longer time than even the hardiest plants and animals could bear. The effect on food chains was appalling, especially for dinosaurs, the comparatively large and hungry creatures at the top of the chain.

Traces of the impact are hard to see because too much has happened to the Earth's surface across that huge area in the last 65 million years. The crater walls have long since been obscured by erosion. But the deepest layers of crust were compacted by the original impact as if they'd been struck by an enormous hammer. Concentric rings of unusually dense rock, centred on the impact zone, affect the local gravity field to this day. A colossal asteroid impact best explains the formation, now known as the **Chicxulub crater**.

There's more evidence of a massive smash in the **K-T boundary**, a thin, ashy layer of sediment found almost everywhere around the world, and dating from 65.5 million years ago. The best explanation for this layer is the gradual settling of dust over the years following the impact. The K-T boundary contains traces of the metal iridium, which is often found in asteroids, but is rare on Earth. Furthermore, quartz crystals in the K-T layer have been deformed by sudden high pressures and temperatures, as might be expected after a colossal impact.

The asteroid may also have triggered a secondary environmental upset: planet-wide volcanism. Vast plumes of gas erupting from fissures in the crust would have made the atmosphere distinctly unpleasant. Whatever the exact circumstances, there are thousands of different dinosaur fossils buried below the K-T boundary, but almost none have been discovered

North Sea impact crater

Phil Allen, a geophysicist based in Scotland, discovered a hitherto unknown meteorite impact crater by pure chance. In the summer of 2002, oil company BP (British Petroleum) asked him to look at 3D computer interpretations of seismic data from a gas field 4km below the North Sea. During his analysis, Allen discovered unusual features in layers of chalk lying above the gas field. A 3km-wide deformation in the chalk looked like a crater, but Allen didn't quite believe it until he met with Simon Stewart, a BP geologist who confirmed his findings.

The crater, named Silverpit in honour of a nearby sea-floor trench, was gouged out by a 200m-wide asteroid that crash-landed around 65 million years ago, suspiciously near the time of the much larger impact that killed off the dinosaurs. Silverpit's impactor may have been a fragmentary accomplice to the greater disaster.

above it. This strongly suggests that, in geological terms at least, something swift and disastrous happened to the dinosaurs some 65 million years ago.

Stones that fall from the sky

Meteorites, the small (and occasionally not so small) rocks that fall from the sky and hit the Earth, originate mainly from the asteroid belt. The asteroid that did so much damage to Earth 65 million years ago is classed as a meteorite, because it made contact with the ground. The little dusty fragments and pebbles that burn up harmlessly in the upper atmosphere are called meteors.

Widmanstätten patterns

Some meteorites are made almost entirely of iron, suggesting they originated in an ancient planetesimal with an iron-rich molten interior. Then, in the cold of space, the small blobs of iron – presumably scattered by some violent disruption – cooled and solidified. Meteoritic iron has a distinct crystalline structure, the Widmanstätten pattern, which can easily be seen when the iron is sliced and polished. For those of us with an imaginative mindset, the pattern looks uncannily artificial, but this is an illusion.

If you can get hold of a small slice of iron meteorite from a reputable dealer, you'll have in your hand material dating from the birth of the solar system. That beguiling metal matrix existed even before Earth itself was born. Iron on Earth has been melted, oxidized and messed about with so much that the Widmanstätten pattern can only be found in those virginal meteoritic samples.

Stony-irons and chondrites

The next most common form of meteorites are stony-irons, which are derived from the shattered inner crust of destroyed planetesimals. Finally, the most common kind of meteorites are stony, very rich in silicate minerals, and contain only small flecks of metal. These originally came from the outermost crust of planetesimals. Many of these stony meteorites are called "chondrites" because they contain **chondrules**, tiny fragments of magnesium, iron, silicon and oxygen compounds fused into distinctive glassy spheres.

The chondrules were formed near the young sun's surrounding accretion disc of gas and dust, and then became incorporated into planetesimals and planets. Earth has been so drastically volcanized, smashed into, melted and remelted by geologic activity that any traces of original chondrule material have long since been obliterated. Chondritic meteorites show us the oldest examples of solid material in the solar system that we have ever encountered.

Carbonaceous chondrites

The most fascinating meteorites are carbonaceous chondrites, so called because they include carbon in some of their mineral compounds. The carbon is bound up with sulphur, iron, calcium, magnesium and other elements in the form of acid derived carbonate salts.

A carbonate salt is produced when carbonic acid (H_2CO_3) acts on surrounding metallic minerals. On contact with metal, the hydrogen atoms in a typical acid become dissociated. Metal atoms replace hydrogen in the overall compound to form a "salt". As far as we know, carbonic acid cannot occur in nature as a pure substance, but it is commonly found as a very weak solution, produced when carbon dioxide gas dissolves in water. Rainwater, in fact, is weak carbonic acid. The more carbon dioxide we pump into the air during industrial fossil fuel combustion, the more "acid rain" we create, but even in the absence of our industries, carbonic acid is natural and commonplace.

The presence of carbonate salts in carbonaceous chondrite meteorites suggests that similarly acidic water may have flowed through their cracks and fissures, presumably when the meteorites were still bound up in the greater mass of their original source bodies. Similar carbonates can also be produced by the metabolism of certain types of terrestrial bacteria. They live in the fissures of rocks, or in sediments, and are sustained by moisture, creating chemical reactions with calcium, manganese, iron and sulphur.

These amazing parallels explain why carbonaceous chondrites are so highly prized among scientists. It is extremely controversial to suggest that any of the carbonates in a meteorite might be associated with living chemistry. At the same time, it is hard to dismiss the action of water on the meteorite's material at some time in the distant past.

Complex chemicals in meteorites

Many carbonaceous chondrites also include organic compounds. Dark, tar-like patches of material contain polycyclic aromatic hydrocarbons, or PAHs. This technical term describes a broad class of compounds consisting of long repeating (polycyclic) chains of hydrocarbon molecules. PAHs are often found in close conjunction with the mineral carbonates inside carbonaceous chondrites. This doesn't necessarily mean that the carbonates and PAHs were formed at the same time. It's just that the tar-like PAHs happen to favour coalescing around the carbonates.

It's tempting to make the imaginative leap from organic chemicals to full-on life, or at least, its tar-like remains. In fact the presence of PAHs in carbonaceous chondrites is so familiar to meteorite specialists that they almost take it for granted. Many scientists believe that organic compounds are created in space by the same kinds of ultraviolet interactions that typically, and so easily, generate the classic Miller–Urey products (see p.36). The solar system's original cloud of dust and gas certainly contained all the right ingredients.

That said, PAHs are often associated with life on Earth, as they're what we expect to find after biological entities have decayed. They are also found in crude oil and car engine fumes. We would never think of oil as directly biological, but we don't call it a "fossil" fuel for nothing. Crude oil is derived from the Carboniferous period, a span of some seventy million years during which the organic detritus from widespread forests and swamplands was compacted, heated and pressurized by geothermal morphism (see p.84). The PAHs that emerge from a car's exhaust pipe are far removed from the original biological source.

Meteorites and life on Earth?

No certain link with life can be presumed unless all the other links in the argument are firmly in place, so we mustn't get too excited by PAHs in meteorites. On the other hand, we can't rule out the possibility that meteorites may have contributed organic material to the Earth. There are about 2500 medium-to-large meteorites in various geological collections around the world, yet they represent only an insignificant proportion of the material from space that has actually hit us. The Earth is constantly bombarded by meteorites so small that we seldom recover

The Murchison meteorite

Hundreds of organic compounds have been identified within the Murchison meteorite, a large carbonaceous chondrite that fell to Earth in 1969 and was recovered in Australia for scientific analysis. These include PAHs and a number of compounds more closely related to the molecules actually exploited by living entities.

In fact, the most intriguing of the Murchison compounds aren't merely similar to fragments of life chemistry, they are utterly indistinguishable. These are aminos, purines and pyrimidines, some of the building blocks of proteins. They're evidence that the Miller–Urey synthesis of aminos and other basic organic compounds happens not just on Earth but also on a grander scale, among the dust and ice clouds of interstellar space.

them, because they are extremely hard to distinguish from ordinary stone debris. At least a hundred tons of these meteoric dust fragments land on our planet every day. As they come to rest and lose themselves among piles of dust and rubble, or sink into the vast oceans, we seldom notice these swarms of alien arrivals. We cannot yet judge to what extent any organic material lodged inside them, or within their larger meteoritic cousins, may or may not have contributed to the history of life on Earth.

It's not impossible that some of the carbonates and PAHs in carbonaceous chondrites were produced by ancient extraterrestrial life, but it's extremely unlikely. Merely accounting for the acidic water that probably formed the carbonates is a tricky problem. Since we know so little about where carbonaceous meteorites came from, we're not really equipped to explain how water could have flowed through them at some time in the dim and distant past. We speculate that most meteorites are fragments of ancient planetesimals, small, half-formed precursors to full-fledged planets. We know so little about those primordial worlds, however, precisely because they've all been smashed into meteorites.

Wouldn't it be much more interesting if we could find a carbon-rich meteorite crammed full of carbonate salts and PAHs that came from a full-sized planet that still exists, one with an atmosphere and, best of all, some water? Then the biological hints and teases inside such a meteorite might not be so easy to dismiss. It just so happens that we've found one. Several, actually, but one in particular is now the most famous meteorite in the world.

Meteorites from Mars

The acronym "SNC" honours the names of three particularly interesting meteorite impact sites: Shergotty, in India; Nakhla, in Egypt; and Chassigny, in France. Just over thirty SNC or "Snick" meteorites are logged in collections around the world, including six found in the Antarctic which have been grouped for convenience into the same class.

The **SNC meteorites** are special, because they don't fit exactly into the stony, iron or stony-iron categories. They don't contain chondrules, but they do show wide variations in age, and other classic signs of geothermal morphism, the shape-shifting changes that happen to rock when it is melted, remelted and generally transformed by the immense forces within a planetary crust. SNCs are igneous – the products of molten rock flowing from beneath the surface crust and making its way through fissures in crustal rock as magma, or bursting right out into the open as lava. Anything igneous is more likely to have been created on a full-sized planet, rather than a planetesimal.

Does the fact that SNCs are not all of the same age suggest that they come from different sources? Not necessarily. If the SNCs come from one source world that was geothermally active, then its crust would have produced rocks of varying ages. Earth provides plenty of examples. The Itsaq rocks of Greenland are over three billion years old, but a number of volcanoes around the world are producing fresh igneous crust at this very moment. Therefore, the SNCs could range widely in age, yet still come from the same source.

The next step in solving the SNC mystery was to explain how they could have been propelled into space prior to their long drift towards an intersection with the Earth. Volcanic eruptions might hurl rocks off a planet, given sufficient explosive power, but it's a stretch. A big meteorite strike could do it, though.

The hunt for the SNCs' home world continues by a process of elimination. The solar system has a gravity gradient which favours objects that fall sunwards. A rock fragment smashed away from Mercury would have to work against the sun's gravity in order to reach Earth. The thick atmosphere of Venus would impede a hurtling fragment on its way up so much that it wouldn't even escape the planet. Among the outer planets, Jupiter and Saturn are not among the suspects. They are gas giants, not worlds made of solid rock. Jupiter's sixteen major moons and Saturn's eighteen are possible candidates, but it seems unlikely that fragments from any of these sources could have escaped the immense gravity fields of their parent planets.

Could the SNCs have fallen to Earth from our moon? Certainly our nearest companion in space has been severely pummelled by meteorite collisions, and has been volcanic in the past. The moon is rich in igneous material. We do find plenty of meteorites on Earth that appear to be lunar in origin, but the SNCs don't match any samples of rock retrieved by the Apollo astronauts in the 1960s and early 1970s during the lunar landing missions.

Clinching evidence for the SNCs' Martian origin came in 1979, when Donald Bogard at NASA's Johnson Space Center in Texas studied meteorite EETA79001, recovered from Antarctica that year. He determined that pockets of carbon dioxide gas trapped inside it possessed carbon and oxygen isotope profiles exactly similar to that of the Martian atmosphere. Why was he so sure? Because the Viking robot landers had made atmospheric measurements on Mars three years earlier. Case closed.

Meteorite ALH84001

The video tape scenes from 27 December 1984 look like holiday mementos. An excitable group of people is driving across an Antarctic ice field in a little powered sled. They are obviously having fun. The sled stops and a parka-clad young woman gets out. The camera's viewpoint jerks downwards to focus on a small dark lump, the size of a potato, lying half-buried in the ice. "Hey, this looks like a good one!" the young woman shouts.

More than a decade after these cheerful keepsake shots were taken, Roberta Score, a member of a National Science Foundation meteorite hunting team, found herself in front of dozens of cameras in Washington, DC, explaining what she had discovered in the remote Allen Hills ice fields of Antarctica. Why had she stopped the sled and instantly noticed that rock? "The colours looked different", she said. "The rock looked very green. It stood out in my mind that it was kind of weird."

> **"Rock ALH84001 speaks to us across all those billions of years and millions of miles. It speaks of the possibility of life."**
>
> President Bill Clinton, August 1996

Just how weird was not immediately apparent to the survey team. Meteorite ALH84001 was logged as a probable asteroid sample. It was sealed into a special canister and shipped to the Johnson Space Center (JSC) in Houston, Texas. For nine years it sat in a protective storage facility, surrounded by nitrogen gas, on a shelf alongside many other similar-looking bits of space rock.

ALH84001's origins

In 1993, NASA geologist David Mittlefehldt requested samples of ALH84001 together with several other meteorites from the JSC collection. He was researching possible links with a very large asteroid called 4 Vesta, the second largest object in the asteroid belt. (Ceres is the largest.) Slicing into his sample of ALH84001, Mittlefehldt found that it wasn't a good match for the suspected Vesta fragments. In fact, he doubted that this meteorite came from anywhere within the asteroid belt. He thought it was more likely to be an SNC. He also noticed some interesting orange patches in the rock. By February 1994, word had spread that an unusual meteorite had been identified, and that it was almost certainly not from an asteroid.

David McKay, chief scientist for astrobiology at the Johnson Space Center, had a particular interest in the ancient Martian environment. After his proposal to study ALH84001 was accepted, he and JSC colleague Everett K. Gibson were allocated a two-gram sample of the meteorite. For six months, an examination at conventional microscopic scales revealed nothing of remarkable interest. Realizing that a more intensive study was required, the JSC scientists joined forces with

Martian orphan and celebrity meteorite ALH84001.

Kathie Thomas-Keprta from the Lockheed Martin company. The new team switched to more powerful instruments (electron microscopes) with far greater magnification. These had become available only a short while before.

By the summer of 1995, it was clear that the NASA scientists were growing very secretive about their work on the meteorite. Naturally, this gave rise to rumours at the Johnson Space Center that they had discovered something of exceptional interest. Those who were not in on the secret had to wait a year before they found out what it was.

On 5 August 1996, the journal *Space News* published its usual reports on NASA's activities, rocket launch schedules and hardware. A snippet by reporter Leonard David picked up on the whispers circulating among the space community: "Meteorite Find Incites Speculation on Mars Life". The story was taken up by the national media that same night.

The next day, NASA's chief administrator Daniel Goldin was forced, ahead of schedule, to put out a formal press release:

> "NASA has made a startling discovery that points to the possibil-
> ity that a primitive form of microscopic life may have existed on
> Mars more than three billion years ago. The research is based on a
> sophisticated examination of an ancient meteorite that landed on
> Earth 13,000 years ago."

Goldin was at pains to point out that all the findings were tentative, and that more work needed to be done. "The evidence is exciting, even compelling, but not conclusive", he warned.

On 8 August, President Bill Clinton, then on the campaign trail seeking re-election, was preparing for a visit to California. NASA's sensational news was a boon to him. Nothing focuses more attention on a president bidding for a second term than appearing on national TV to announce a great American achievement. "If this discovery is confirmed", Clinton told viewers, "it will surely be one of the most stunning insights into our world that science has ever uncovered."

The long space odyssey of ALH84001

Approximately fifteen million years ago, ALH84001 was thrown into space by another meteorite, probably from the asteroid belt, as it smashed into the surface of Mars. The evidence for this is twofold. Firstly, ALH84001 bears the traces of a rapid and powerful impact, in the form of narrow fractures and other shock features associated with SNCs. Secondly, the outer layer of the rock shows signs of exposure to cosmic rays, the power-ful, ever-present radiation that pervades all the galaxy. The impact of these rays on the wandering Martian orphan gradually created a distinct kind of radioactive clock. By measuring the final dosage, geologists have timed the length of the rock's exposure to raw space, which comes out at around fifteen million years and change.

Having been hurled into the Martian sky and then far beyond its thin boundaries, ALH84001 escaped the gravitational pull of its home world and orbited the sun, just like an asteroid. However, most of the asteroids are shepherded into their main belt by the gravity field of Jupiter. ALH84001 was not amenable to this kind of control. It probably embarked on its journey through space on a more random trajectory. We can't be sure how often it may or may not have intersected the Earth's orbit. What we do know is that eventually it passed close enough to be captured and pulled towards the ground.

It fell through the atmosphere and landed in the Allen Hills region of the Antarctic. Heated by its fiery atmospheric entry, it melted the ice that it thwacked into, burying itself deep below the surface. After a short time, the meltwater around it froze once again, sealing the meteorite in place and preserving it against surface weathering.

We can be reasonably sure that ALH84001 remained locked in the ice undisturbed for around 13,000 years, until wind erosion and other com-plex movements of the ice exposed it once again, enabling its discovery. The evidence to support this comes from the fact that ice sheet structure is laminated, rather like sedimentary rock. Certain ancient geological events, such as major volcanic eruptions, have given particular layers their own datable characteristics, adding minute traces of chemicals and dust. (Scientists have noted, with sadness, how the recent arrival of global industrial pollution is clearly indicated by a marked dirtying of ice layers younger than three hundred years.)

There is no particular reason why the Antarctic should attract more meteorite falls than other parts of the Earth. As we noted earlier, there must be countless small meteorites littering the world, disguised to all but the most expert eye as ordinary pieces of terrestrial rubble. They're difficult to identify unless they've gouged out clearly identifiable impact craters. The main reason why meteorite hunters adore the Antarctic is that mysterious isolated rocks tend to stand out against the dazzling white panorama. When a dark stone sits among otherwise pristine sheets of ice, far away from any obvious source, it's a safe bet that it must have dropped from the sky.

Clocking a space rock

Rocks of all kinds tend to contain tiny traces of radioactive elements dating from the origins of the solar system, such as the metallic element rubidium. The radioactive rubidium-87 isotope has a half-life of 48 billion years. Half a given population of rubidium atoms will decay over that period, yielding a different metal, a stable isotope of strontium. The ratios of rubidium to strontium in a rock reveal the time that has elapsed since its last solidification, give or take a few hundred million years. According to rubidium–strontium data, the SNCs appear to be younger than the supposedly asteroidal meteorites. This increases the likelihood that they were created on a full-scale planet.

Inside the rock

McKay and Gibson examined the curious orange patches inside ALH84001, each about the size of a grain of sand, which are clearly visible under an ordinary optical light microscope. They occur on the walls of fractures inside the meteorite. At first glance these carbon-rich features, which McKay described as "globules", look more like the sort of deposit often left on rock surfaces by terrestrial anaerobic (oxygen-averse) bacteria. The inner cores of the globules are rich in calcium and manganese. Cross-sectional cutting revealed a surrounding band of iron carbonate, while the outer layers consist of iron sulfides. This banding of different materials within the carbonate structures suggested a more complex chemistry than non-living chemistry alone could achieve.

In addition, tiny grains of **magnetite**, an iron oxide, were found inside the carbonate globules. Specks of magnetite are created in the bodies of bacteria which live around rocks, or in silts and sediments. The mineral appears to have some benefit, allowing bacteria to align themselves advantageously in relation to their environment. The grains of magnetite within ALH84001 have surfaces suggestive of biology. Gibson and McKay argued that the lack of sharp crystal faces on these surfaces was indicative of a bacterial source. Critics argued that Martian bacteria couldn't have benefited from magnetite, for the stark and simple reason that Mars has a very weak magnetic field. On the face of it, there is no obvious reason why any native organisms should have evolved these little magnetic "compasses". However, billions of years ago, the planet's iron-rich interior had not yet solidified. The magnetic field was probably stronger in the distant past than it is today, because the "dynamo" effect of the iron core's different rate of rotation from the outer crust would still have been evident.

This interpretation of the grains was controversial, to put it mildly. As with all the individual features found within ALH84001, the grains could also have been created by non-living processes. In December 1996, a team led by John Bradley at the Georgia Institute of Technology, investigating their own little shard of the meteorite, managed to slice through one of the magnetite grains, using an intense beam of ionized subatomic particles. The team found a kind of internal corkscrew effect in the structure which, they said, is not normally found in bacterial magnetite. As always, McKay and Gibson had an answer. There was too little data on earthly bug magnetite, they contended, for Bradley to be sure that corkscrew magnetite wasn't biological.

Too small to live?

The most controversial and by now globally familiar piece of evidence wasn't highlighted in the initial paper for *Science*. It seemed altogether too fantastic. At the extreme magnifications delivered by their scanning electron microscope, McKay's team picked out physical structures resembling **fossilized bacteria**. They were teardrop-shaped, which was unusual in itself, as if some deliberate gradient of structure were present. Some of the examples even appeared to be segmented, like tiny worms.

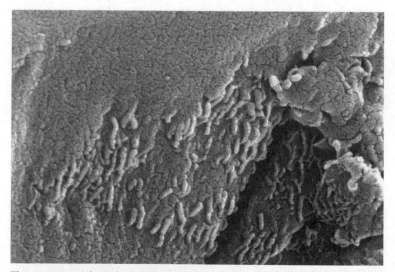

The controversial teardrop-shaped "fossils" in meteorite ALH84001, which some scientists believe are traces of Martian life.

The trouble is, these "fossils" were smaller than any bacteria yet found on Earth. Carl Woese at the University of Illinois came up with a further problem, telling the journal *Meteoritics and Planetary Science* that "these putative Martian bacteria are scarcely larger than the magnetite grains they're supposed to have produced." He said that the bug-like features revealed by the NASA team's electron microscope beams were so small, they were more likely to be accidental products of mineral deposition processes than the remnants of living organisms. "At that tiny scale, it's easy to read significance into random formations", he warned.

But who's to say how small a living entity can be? In the same month that McKay and his colleagues announced their findings from the meteorite investigation, Todd Stevens of the Pacific Northwest Laboratory in Washington State announced the discovery of extremely small bacteria living in rock fractures deep beneath the Columbia River. These appeared to be dependent on mineral reactions for their energy supplies, and were just twice the size of the supposed fossils in ALH84001.

Six years later, at the New Mexico Institute of Mining and Technology, Thomas Kieft identified what he calls "dwarf bacteria". These are starved, semi-dormant organisms in sediments that have become hardened and compacted because of drought and subsequent shrinkage. So there's no reason why Martian bugs inside ALH84001 shouldn't have produced magnetite grains as full-sized, healthy bugs, and then become shrivelled and dormant as circumstances changed – perhaps when the Martian environment altered for the worse and the water disappeared. The supposed fossils in ALH84001 might represent dried-up versions of once-healthy microbes.

University of Texas geologist Robert Folk is a leading proponent of nanobacteria on Earth. He argues that "they are enormously abundant in minerals and rocks, and probably run most of the Earth's surface chemistry". Until the 1990s, many scientists assumed that bacteria could only shrink to a certain point, a limit that happened to coincide with the resolving power of their best instruments. Folk and many other researchers now think that we have to set aside these old assumptions. Biological activity could account for a major proportion of what we're accustomed to describing as mineral processes in the Earth's crust. We may even have to rethink our understanding of crude oil deposits.

"If nanobacteria are so abundant – possibly an order of magnitude more abundant than normal bacteria – how is it that they could have been missed for so long?" Folk asks. "Probably the main reason is that microbiologists have little or no interest in the occurrence of any type of bacteria in soils or rocks, and it has been standard dogma for fifty years that bacteria smaller than a certain size cannot exist." If Folk and his supporters are proven correct, then the nanofossils in ALH84001 will look even more convincing. Even though the status of nanobacteria as living entities is hotly debated, it looks as if we'll have to reinvent our explanations for a good deal of terrestrial geology too.

Are there really fossils in ALH84001?

The debate rumbles on. So far, the pro-life explanations accounting for the carbonates and PAHs in the Mars meteorite have not gained an absolute victory over the non-life possibilities. In theory, the carbonates, the magnetite grains and even the strange teardrop-shaped "fossil" features could have been produced entirely by inorganic chemical reactions driven by acidic water flowing through the cracks in ALH84001 while it was still a part of Mars. However, in their paper for *Science*, McKay and his colleagues argued that varying levels of acidity and alkalinity (pH levels) needed to prevail simultaneously in order to produce the differentiated materials that they found. (A hallmark of biological metabolism is its ability to sustain a range of pH levels within the same system.) "Any one of our findings could be a product of mineral reactions", he told a news conference at the time, "but the separate indications that we found, all in close proximity, are much harder to account for without a biological explanation. You have to explain all these things together, not just in isolation. In the end, the biological model is just so much more straightforward than all the others."

The challenge of proof works both ways. The scientists who disagree with McKay and Gibson have an equally tough time making their arguments stick. The relevant scientific techniques are infuriatingly subtle, and so many strands of evidence overlap. Conflicting, yet equally plausible "explanations" for the observed data can be – indeed, have been – proposed. It's not really reasonable to expect more definitive evidence from ALH84001. After all, the meteorite is billions of years old, and has been smashed into by another meteorite. It's drifted through space for millions of years in extremes of vacuum, cold, and exposure to radiation; it's slammed into Earth's atmosphere at five times the speed of sound; and it's been exposed to 13,000 consecutive winters in one of the coldest and most inhospitable environments in the world, the Antarctic. These circumstances are not suited to the preservation of delicate biological clues. For those, we'll have to go to Mars and obtain some fresh samples.

Will humans ever get to Mars?

In July 1989, US president George Bush Sr celebrated the twentieth anniversary of the Apollo 11 moon landing. With Neil Armstrong, Buzz Aldrin and Mike Collins standing at his side, he talked of "a journey into tomorrow, a journey to another planet, a manned mission to Mars". Short of setting a precise target date, he seemed to have given the green light for the boldest rocket project since John F. Kennedy's 1961 speech directing the US towards the moon "before this decade is out".

NASA set to work on the infamous "90-Day Study on Human Exploration of the Moon and Mars" (more commonly known as the **90-Day Study**), so-called because that was how long it took to deliver a plan almost guaranteed to make Bush wish he'd never mentioned Mars. A thousand-ton interplanetary craft was to be assembled in Earth orbit. The Mars return trip would take eighteen months, with just a few weeks spent on the surface: enough time to plant a flag and snap some photos before heading home. NASA priced the project at $200 billion, at which point everyone apart from the space cadets lost interest.

At the Martin Marietta aerospace company, engineer Robert Zubrin was appalled by these costings. It seemed to him that building a giant spaceship had become more important than the Mars mission itself. Al

President George Bush Sr announces an ambitious "manned mission to Mars", something he was bound to regret once he heard the estimated cost.

Schallenmuller, Marietta's chief of civilian space systems at that time, fondly remembered building the Viking landers in the early 1970s, and was keen to see a human Mars mission if it was at all possible. He allowed Zubrin and a dozen colleagues time to rewrite the company's sales pitch to NASA. By February 1990, the team had cut the weight of the outbound ship in half and slashed the costs. Their "Mars Direct" proposal is now the best hope for the future.

Mars Direct

Most Mars proposals call for a huge mother ship to circle the planet and send down small landing teams, which then come up, rendezvous with the ship and fly home. Zubrin calls this the *Battlestar Galactica* approach. Zubrin asks, why have the mother ship at all? In Mars Direct, you fly hardware directly from Earth to Mars, and then a small Earth Return Vehicle (ERV) fires off the surface and heads back home. This smaller style of ship would mean confining astronauts to relatively cramped cabins for the six-month outward and eight-month return trips, but as Zubrin points out, "we know from the International Space Station experience that people can tolerate that if they're sufficiently motivated. We don't have to build giant space cruisers to go to Mars."

> "Landing humans on Mars requires neither miraculous new technologies nor vast expenditures. We simply need to use common sense, and employ the technologies we have at hand now."
>
> Robert Zubrin, *The Case for Mars* (1996)

A huge amount of fuel (and therefore spacecraft size and weight) is saved by launching only when Earth and Mars swing close to each other in their orbits and are both on the same side of the sun at once. This happens roughly every two years, so Mars Direct employs a rolling schedule of missions to coincide with these close planetary approaches. The downside to this low-energy scheme is that each mission takes well over two years from start to finish, since the crew have to wait on Mars for several months until the planet swings close to Earth again and they can begin the return trip. On the face of it, this lengthens their exposure to cosmic radiation hazards. But while Mars Direct involves a longer stay on the planet, astronauts actually spend less time in deep space, where the radiation hazard is greatest.

Supporters of Mars Direct also point out that space radiation, the supposed game-closer for human deep space missions, doesn't kill people. What it actually does is increase the risk of radiation-related cancers in

Living on Mars... on Earth

Devon Island in the arctic Nunavut Territory of Northern Canada is bleak, cold and dry. There's a dramatic 20km wide crater here, the 23-million-year-old product of a meteorite strike. In 1997, NASA scientist Pascal Lee set up a project to explore the crater. When he saw it for the first time, he knew he had found the perfect place for training future Mars explorers. In conjunction with the Mars Society (a 5000-strong group of scientists, engineers and Mars enthusiasts), Pascal raised more than a million dollars in sponsorship from the Discovery Channel and Flashline software company, for the construction of a replica Mars habitat on Devon Island.

Life at the Flashline Mars Arctic Research Station (FMARS) is one of the strangest games of make-believe that anyone ever played. Teams of scientists are learning how to live and work on the Red Planet. The rules are strict. No one goes outside the habitat module unless they're wearing a space suit. Then, on returning, the suits have to be vacuum-cleaned to get rid of all the dust from the outside, as if guarding against genuinely corrosive Mars dust.

The FMARS crew communicate by radio with NASA's Ames Research Laboratory in California. No one is allowed to have a normal, "real-time" conversation, because in a real Mars mission, radio signals take, on average, ten minutes to travel to Earth. An artificial time delay is built into the radio link.

Russia and the European Space Agency (ESA) are collaborating on a similar Mars simulation project, called Mars500, at the Institute of Biomedical Problems (IBMP) in Moscow. The volunteers (engineers, doctors and scientists from Russia, Europe and China) are sealed up in a mock spacecraft and live in the same cramped, isolated and minimally stocked conditions that their cosmonaut counterparts would experience in space, simulating the psychological challenges of a five hundred-day Mars mission without leaving the ground.

Not an astronaut hoovering in his spacesuit, but a Mars500 crewmember doing simulated scientific research on mock Martian terrain.

later life. So far, the overall death rate of astronauts shows that (accidents aside) their longevity is no different from normal. Weightlessness is another supposed problem. We know that astronauts on extended space missions lose body mass, muscle tone and bone strength. Ever since the dawn of the Space Age, engineers have dreamed of vast wheel-shaped stations whose rotation creates a centrifugal force, pressing the human occupants' feet against the inside surface of the wheel's outer wall, creating a sensation similar to gravity – or so the theory goes. A famous scene in Stanley Kubrick's 1968 film *2001: A Space Odyssey* makes such visions seem compellingly achievable. In practice, weightlessness is exactly the condition that justifies building a space station in the first place, because this enables unusual scientific experiments to be conducted on board.

In the case of a trip to Mars, some kind of **artificial gravity** might be desirable, although no one would have the budget for building a giant wheel for the living quarters. Rather, the ship would look more like a tumbling stick. Think of the stick as two spokes radiating in exactly opposite directions from the centre of an imaginary wheel. Now cut away most of the wheel's rim, except for two stubby crew compartments, one at each end of the spinning stick. The centrifugal principle is the same, and the astronauts would experience a feeling of being pressed against the floors of those two outer compartments.

Making fuel from the Martian atmosphere

Landing modules could arrive on Mars with their fuel tanks half-empty, thus saving even more weight and reducing the cost of launching them from Earth. An In-Situ Resource Utilization (ISRU) plant could pump Mars's carbon dioxide atmosphere through a nickel catalyst, adding a trace of hydrogen into the mixing chamber. The catalyst splits the carbon dioxide, liberates the oxygen and combines it with the hydrogen to make water. The freed carbon reacts with spare hydrogen to create methane, which would serve as rocket fuel for the return trip. Six tons of hydrogen carried to Mars could be converted into one hundred tons of methane and oxygen.

The methane is pumped straight into the return vehicle's tanks. Meanwhile, a weak electric current passes through the water to extract the oxygen, while the water's hydrogen, now free again, is pumped back into the system so that the cycle can begin anew. This is a simple electrolysis process, familiar to us all from science lessons. Some drinking water could be retained, but in the main, astronauts would recycle their urine and sweat for that. No new technology has to be invented to make ISRU work, and the technique has already been proven. Future robotic Mars probes will carry miniature ISRUs.

The astronauts would reach Mars in pretty good shape, yet without having to travel in a huge, complicated and expensive ship. For the return voyage only one module would be used. The second module and the stick-like linking tether, with its gift of artificial gravity, would be lost. If, as a result, the astronauts were a little less than super-strong on their return, it wouldn't matter. There'd be reception teams and plenty of support on standby. Mars explorers would land on Earth in no more discomfort than astronauts coming home after long missions aboard space stations.

There is always a chance that a crew might get stranded. What if their Mars lander crashed, or failed to take off at the end of the mission? Again, this is a simple problem to solve. Uncrewed return vehicles can be sent to Mars under computer control. The modules then touch down and report back their status. Only when they are sitting safe on the surface, ready to lift off again when required, would humans be sent to join them. They'd need to touch down within reasonable walking (or wheeled roving) distance of the waiting return ships. The key point is that no one need be sent to Mars without being assured of a trip back home. If the first fleet of robotic ships crashed, the entire human mission would be delayed for a few years until the next set of "empty" return vehicles could be sent.

Political hurdles

One by one, Zubrin and his army of supporters in the highly organized Mars Society lobbying group are trying to knock down the anxieties and cost implications that, so far, have prevented us from sending people to Mars. NASA is now listening. The Mars Society has also attracted corporate sponsorship to build trial Mars habitats in some of Earth's hottest, driest and coldest locations (see box on p.95). Teams have already experimented with the disciplines relevant to a real mission.

In January 2003, President George W. Bush announced that Mars should be the destination for a new Vision for Space Exploration as a logi-cal next step from NASA's venerable winged Earth-orbiting space shuttles. His advisors were determined not to repeat the mistakes of the 90-Day Study. The first human footprint on Mars seemed to come one small step closer, until a new president, Barack Obama, reapplied the brakes. It's not technology, so much as dull stuff like politics and lack of cash, that blocks the road to Mars.

If not Mars, where?

The discovery of microorganisms living in some of the strangest and most extreme environments on Earth has caused us to reassess the range of worlds within the solar system where life might be found.

At the dawn of our solar system, the sun was not the only competitor for dust and gas in the surrounding accretion disc. The massive hydrogen-rich gas giant that we now know as Jupiter could have become a star in its own right. The sun absorbed the lion's share of hydrogen and helium from the cloud of gas and dust until the supply simply ran out. Jupiter did not quite achieve the high temperatures and pressures at its core that a true star needs. Had it been able to sweep up materials from the disc for a little longer, growing to just seven times its current size, it could have burst into

The Jovian system

Jupiter is the fastest-spinning planet in the solar system, taking less than ten hours to rotate around its axis. The rapid rotation causes Jupiter's equator to bulge outwards. Rotational energy also shapes the planet's atmosphere, cloud belts and storms into distinct bands and whorls. Jupiter's famous Great Red Spot storm system was first observed by astronomers more than three hundred years ago, and shows no signs of abating today.

Jupiter has sixteen significant moons. Io, the third largest, is in permanent volcanic disarray, its surface stained with yellow sulphurous outpourings. The tidal forces generated by Jupiter's gravity are so great that Io's crust can never settle. Ganymede is Jupiter's largest moon – in fact it's the largest in the entire solar system – and the only moon with its own magnetic field. This suggests that it has a molten iron-rich core. Callisto, Jupiter's second-largest moon, is about the same size as Mercury. It orbits just beyond the reach of Jupiter's fierce radiation belt, and is the most heavily cratered satellite in the solar system.

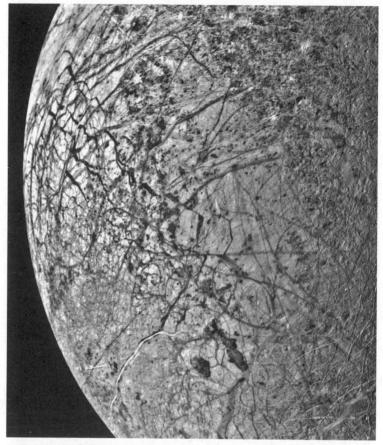

Europa may harbour a vast body of water beneath its fractured icy crust. Where there's water, there could be life.

light and become the junior partner in a binary star system. Luckily for us, this never happened. Binary star systems, with their double helping of ultraviolet radiation sources and chaotic gravitational whirlpools, are not ideal places for life-bearing planets to form.

A hidden ocean

Europa, Jupiter's fourth largest moon, is especially fascinating. In August 1996, the Galileo space probe beamed back detailed images of its icy surface.

The crust is an uncannily close analogue for earthly Arctic and Antarctic seas in deep winter. Jagged polygons of surface material on Europa appear very like chunks of pack ice in arctic oceans that drift around and split into smaller chunks, freezing into place when the surrounding ocean waters also harden into ice during the winter. Scientists think that a huge ocean of fluid water exists just beneath Europa's frigid crust.

Close-up views of the crustal fractures reveal many parallel strips, with bright runs of icy material flanking dirtier-looking bands. It seems as though a dark, viscous material is oozing up in the cracks between the ice floes, leaving murky stains. Whatever slushy stuff exists beneath the topmost layers, it seems to be of a different colour and chemical composition to the cleaner ice on Europa's surface.

Could this indicate Europa's suitability for life? Tidal forces from Jupiter prevent the subsurface water from freezing. Quite possibly, the water is comparatively warm. Even without sunlight, organisms in the hidden ocean could be sustained by heat from hydrothermal vents on the ocean floors. We know this is possible because we've found such organisms here on Earth

Black smokers

In the late 1970s, scientists in the deep-sea submarine *Alvin*, operated by the Woods Hole Oceanographic Institution in Massachusetts, plunged to the bottom of the eastern Pacific, some 2000m below sea level. They found hydrothermal vents, eerie chimney-like structures belching superhot mineral-rich fluids into the cold, dark waters above. Known as **black smokers**, these vents typically form along ridges where tectonic plates diverge. New molten crust is pushed from below into the ragged space that opens up, then it cools and sets into place. The "chimneys" form when dissolved metals suddenly precipitate as the superhot water bursts upwards and meets the surrounding ocean water, which is only a few degrees above freezing. These strange towers spew a noxious brew of chemicals, giving the "smoke" its black colour.

Black smokers seethe with extraordinary, unexpected life such as tube worms, clams and eyeless shrimp. Our familiar food chain on Earth's surface is based on energy from the sun, which is harnessed by plants. The grass turns into cows, the cows become hamburgers, and so on. That life-powering sunlight never reaches the deepest sea floors. Here, organisms must rely on a different energy source: the metals and other chemicals that rocket out of those vents. In one especially intriguing reaction, hydrogen sulphide and iron monosulphide deliver the energy for microbial life to thrive, forming the mineral iron pyrite ("fool's gold") and hydrogen gas. It's hydrogen, rather than sunlight or oxygen, that provides energy for the bacteria at the bottom of the food chain on the floor of the deepest oceans.

All this means that our ideas about the earliest forms of life on Earth may be askew. The problem with the traditional blue-green algae scenario (see p.26) is that the supposedly primitive bugs involved in that process must have been quite complicated, because they would have contained sunlight-processing chlorophyll, a pretty advanced piece of chemistry in its own right. What kind of simpler life might have existed before the algae? Black smokers may hold the key to this riddle. Perhaps life begins in the sunless depths rather than up above? If so, we must widen yet further our ideas about what kind of extraterrestrial environments might make suitable domains for life. Unfortunately, it will be a few years yet before we next have a chance to look at Europa in more detail. ESA and NASA are considering a joint mission that could lift off in about ten years' time.

The liquid lakes of Titan

On 14 January 2005, a tiny European space probe touched down on Titan, a far stranger and even more distant world than Europa. Huygens took seven years to get there, spending most of that time clinging silently to the flanks of NASA's Cassini mother ship as it swept across the solar system on an epic three billion-km journey to explore Saturn and its moons. On arrival, Huygens was only designed to operate for a few hours before losing contact with its mother ship. That was enough time to give us a glimpse of an awe-inspiring alien landscape.

As it plunged through Titan's murky sky underneath its parachute, Huygens confirmed what scientists have long suspected. The atmosphere consists mostly of nitrogen, but there is also a thick smog of methane clouds. When the cameras switched on at 150km altitude, some real surprises were in store. ESA scientists at Darmstadt in Germany knew from the first frames that they had struck gold. As imaging team member Anthony Del Genio commented a few days after the mission's climax, "We were lucky to come down smack over a region where you could see a wide variety of surface types in the same scene. That gave us a lot of perspective." The cameras showed that methane rain falls on the ground, and liquid methane rivers scour Titan's landscape as surely as rain carves the lands of Earth. In Titan's nippy temperatures of -170°C, water freezes hard as rock, while methane becomes a liquid.

But where do those rivers end up? From further out in space, Cassini's infrared camera picked out regions of light and dark surface texture beneath Titan's dense atmospheric haze. Hurtling towards touchdown, Huygens showed that the brighter areas are rough highland terrain, probably consisting of water ice. Dark low-lying zones are flat, and have distinct shorelines.

So it looks as though the methane rivers feed vast lakes. They may not be filled with liquid you could swim in, but scientists are confident that they are lakes of some kind. Martin Tomasko, leader of Huygens' camera team, commented, "We're seeing Earth-like processes with very exotic materials. There's plenty of evidence for flowing fluids."

The sharp colour differences between the high ground, the river channels and lake beds caused a stir. Although the sunlight hitting Titan is barely one percent as intense as the light that strikes Earth, it is nevertheless strong enough to trigger chemical reactions in Titan's upper atmosphere, breaking down methane to create a haze of complex hydro-carbons. These fall to the ground when it rains, leaving dark deposits that are rich in the basic organic building blocks of life.

Before we get too excited, we have to remember there's no flowing water, which biologists insist is essential for life. Titan is unimaginably cold, and the only available oxygen is tied up in solid water ice. It is unlikely that any complex wildlife lives on Titan. Even so, this strange moon is an intriguing place of swift change and flux: a world with not a single crater. This is no dead hunk of rock, but a dynamic powerhouse of Earth-like erosion and weather influences that constantly reshape the landscape, smoothing away any traces of impacts. There is even evidence for volcanoes. Instead of spitting out molten lava and hot gas, they disgorge frigid ammonia and water ice.

Icy wanderers: comets

We often see thin, bright streaks in the sky at certain regular seasons during the year. These momentary flashes are caused when Earth passes through diffuse clouds of dust and ice particles drifting in space: the debris left behind by comet tails. When dust grains from a tail collide with Earth's upper atmosphere at colossal speed, the little specks burn up for a second or so in a last blaze of glory before vanishing.

Two meteors (the two streaks strafing downwards from right to left) of the Perseid meteor shower light up the skies over Jordan in August 2004.

These **meteor showers** are as insubstantial as sprites. Every year astronomers enjoy a particularly well-known display, the Perseid shower, which peaks in intensity around mid-August with faithful regularity. This shower has been traced to a comet which last swept across the Earth's orbit in 1862, leaving a slight memory of its tail in its wake. Eventually the tail will dissipate, and with it, the beautiful Perseids.

Comets appear to be the frozen remains of countless small-scale accretion events during the birth of the solar system. Light and insubstantial, with almost no gravity, they failed to pull in enough additional material to become protoplanets or protostars. Most comets – and there are huge numbers of them – roam the Oort Cloud, named after the Dutch astronomer Jan Hendrik Oort, who predicted its existence half a century ago. A vast spherical zone far beyond the orbit of Pluto, it reaches a third of the distance to the next nearest stars.

Once in a very long while, the sun pulls a stray comet towards the centre of the solar system. It then sweeps a long and lonely elliptical orbit, occasionally passing close enough to Earth for astronomers to take notice. Even more occasionally, a comet passes close enough for us to see its glowing tail in the night sky. Comets that interact with the solar system repeatedly, at regular intervals, drift in a disc-shaped realm known as the Kuiper Belt, beyond the orbit of Neptune, but closer to us than the Oort Cloud.

The main asteroid belt between Mars and Jupiter stays broadly on the flat and level. By comparison, comets seem to travel at all kinds of crazy angles, punching in and out of the solar system's disc-shaped "plane of the

Comets up close

Some comets sweeping through the solar system have been observed by space probes. If these occasional specimens are any reliable guide, comets are irregular lumps of ice and dust ranging from a few metres to tens of kilometres in diameter. The main constituents of the core, or nucleus, are water ice, carbon dioxide ice and silicate dust grains. These loosely compacted snowballs are usually so delicate that solar radiation shreds their outer layers and creates characteristic glowing tails of ionized particles.

Many comets contain simple organic molecules, such as ammonia, formaldehyde, methyl cyanide and other hydrocarbons. Comets falling onto the young Earth probably contributed some of the building blocks of life. Some planetary scientists think that most of the water in our oceans came from cometary ice.

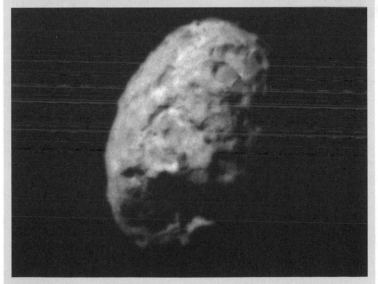

Comet Wild 2, as photographed by NASA's Stardust space probe.

ecliptic" (the zone of the planets' orbits) like kingfishers darting up and down through the surface of a lake. But why should they have escaped the solar system's flat, gently spinning realm? What kind of energy could have disrupted such vast numbers out of the plane?

We don't exactly know where comets originated. It has even been speculated that they are not members of the solar system at all, but vastly more ancient entities, icy little orphans wandering between the stars. They may be the frozen remains of countless accretion clumps which never acquired sufficient mass to become significant celestial bodies. Capturing a comet might be like investigating the miniature egg of a planet, or even a star, that never came into being.

A universe fit for life?

Confrontational and gruff-spoken Yorkshire-born cosmologist **Fred Hoyle** (1915–2001) lost his respect for authority at primary school. A teacher told

> **"There is a coherent plan in the universe, though I don't know what it's a plan for."**
>
> Fred Hoyle (1915–2001), cosmologist

his class that a certain type of flower has five petals. The next day, Fred produced a flower of the same kind with six petals and asked the teacher why she'd said it had five. The teacher clouted his ear. As a young man he displayed an astonishing talent for mathematics, graduating from Cambridge University with the highest marks in his year. But he never trusted authority figures, and spent much of his life arguing with anyone who tried to tell him he was wrong about anything.

In the early 1940s, Hoyle contributed to the British-led development of radar. A secret research trip in 1944 brought him into contact with America's atomic bomb theorists, leading him to consider how nuclear processes inside stars might be responsible for creating chemical elements. In the spring of 1953, Hoyle burst into the office of his good friend Willy Fowler, a nuclear physicist at California Institute of Technology (Caltech), and an expert on carbon atoms. "I exist, and I'm made out of carbon!" Hoyle exclaimed. "Therefore the carbon-12 nucleus must possess an energy level at 7.65 megaelectronvolts (MeV)."

This was his technical shorthand for an amazing idea he'd come up with. Hoyle believed that nuclear reactions inside a star were responsible for fusing hydrogen and helium, the basic building blocks of the universe, into all the heavier elements, such as oxygen, carbon, nitrogen and so on, up to the mass of iron. Anything heavier than iron is created in the final, cataclysmic explosion of a supernova. For his theory to work, carbon inside a star had to exist at that special (and unusual) energy level before it could contribute to the fusion of heavier elements.

Although Fowler would later tell people that his first impression of Hoyle was of "someone who had lost his mind", he listened patiently, then persuaded experimenters at Caltech's Kellogg Radiation Laboratory to test Hoyle's claim by boosting carbon-12 atoms to the specified energies in a particle accelerator. A few days later, the experimenters confirmed that Hoyle was right on the money. What surprised Fowler, even more than the unnerving accuracy of Hoyle's calculations, was that Hoyle had based his argument on what's known as the **Anthropic Principle**. In his interpretation of a multifaceted idea, the universe is biased towards the creation of life, which cannot arise without carbon. There has to be a mechanism in the laws of nature to generate carbon in the right way, and with that special and extremely fine-tuned resonance at 7.65 million electron volts. Otherwise none of the other elements could be formed, and life would not exist. Fowler thought this sounded mystical and unscientific. He and Hoyle struck up a long friendship nevertheless.

Anglo-American physicist Geoffrey Burbidge, and his astrophysicist wife Margaret, took up the cause and worked with Hoyle on calculations that took five years to complete. They and Fowler were eventually honoured with Nobel citations for proving that all chemical elements beyond hydrogen and helium are created inside stars, up to the level of iron. As Hoyle had theorized, a supernova would be required to produce anything heavier. The cliché is true; we are all star stuff. But Hoyle missed out on the Nobel prize. His argumentative streak worked against him, and some of his other ideas were too radical for the health of his scientific credibility, not least his lifelong insistence that the Big Bang theory is nonsense.

Panspermia: life from comets?

In the late 1970s, Hoyle and his Sri Lankan-born student Nalin Chandra Wickramasinghe developed a theory suggesting that planets are "seeded" from space by comets carrying a rich supply of ingredients. This idea is known as **panspermia**. Wickramasinghe was among the first astronomers to prove that interstellar gas clouds (nebulae) are routinely rich in organic materials. He argued that comets and nebulae are intimately related. The presence of relatively intriguing organic molecules throughout interstellar space is accepted as a fact. However, it's not so certain how complex any of those compounds can become.

Wickramasinghe suggests the following scenario: tiny dust grains of iron, silicates, graphite (a crystallized form of carbon) and other inorganic materials drift through space, occasionally bumping into organic molecules synthesized by the effects of ultraviolet starlight interacting with the grains. Simple organic compounds are created on the surfaces of the grains. If they encounter similarly coated grains, the organic molecules will tend to merge and become more complex, especially if this process occurs within pockets and shadowy recesses in dust grains, where the delicate organics avoid further interference from stellar ultraviolet interaction.

Writing for *New Scientist* in April 1977, in support of more formal papers elsewhere, Wickramasinghe asserted that a form of Darwinian evolution takes place within a dust cloud. There is some degree of competition between different varieties of grain clumps, stimulated by their constant creation or destruction in the interstellar environment. "The simplest self-replicable system involving clumps of inorganic grains glued together with organic polymer coatings becomes most widespread in the galaxy", he contended. "Finally, we could imagine that the organic polymer films which separate original grains would evolve into biological cell walls."

Hoyle and Wickramasinghe discussed their joint theory in *Our Place in the Cosmos* (1993), suggesting that if any of these grain–organic clusters happen to become part of a planet, they would provide key sources of material for life. Meanwhile, the grains which accumulate into comets will retain any complexity that they might already have achieved, since they will not be subject to the violent temperatures and pressures of planetary formation. "It would not be unreasonable to suppose that simple life forms developed and flourished within huge aqueous lakes in billions of large comets", they conclude.

Responding to David McKay's speculation that meteorite ALH84001 might contain Martian bacteria (see p.92), Wickramasinghe declared in an August 1996 *New Scientist* interview, "the reason that primitive life on Earth and Mars appear to be similar is that both planets were seeded by similar organisms."

Startling Stardust discoveries

Organic materials of often quite startling complexity probably do rain down on Earth, and all other planets for that matter. But when Hoyle tried to claim that viruses, bacteria and even multicellular organisms might also fall to Earth inside cometary taxis, this was largely dismissed by other scientists.

When comets strike

Comets might seem fragile, but the consequences can be dramatic when a nucleus, rather than just the dust and ice from a tail, strikes the Earth's atmosphere. On 30 June 1908, some Tungus farmers in a remote region of Siberia reported a huge fireball which seemed to "split the sky in two", accompanied by hurricane winds and deafening booms. Around many parts of the world, the night sky glowed with unnatural brightness for two days. Over London, it was so bright you could read a newspaper. But at the time, few people were able to make the connection between those strange skies and the stories told by faraway Tunguskan peasants. What with World War I and the Russian Revolution, it was more than two decades before any scientists were dispatched to investigate the Tunguska rumours.

Astonished investigators found that 2000km² of forest had been completely flattened. In the central region of the catastrophe, thousands of trees were burned to a crisp. The obvious possibility was that **Tunguska** had been struck by a huge meteorite, but for all the damage, there wasn't the slightest trace of a crater, let alone any meteorite fragments. However, if an icy comet collided with Earth in 1908, it could have hit the atmosphere at a velocity of 30km per second, creating the equivalent of a massive nuclear detonation above the ground.

Close examination of the impact site in recent years has revealed countless numbers of microscopic diamonds in the topsoil. Nobody is quite sure why, but it may be a phenomenon connected with intense temperatures generated by the blast. Apart from that, not a trace of Tunguska's destructive visitor has ever been discovered. Although the comet (if that's what it was) must have turned to vapour before it hit the ground, the "air burst" explosion unleashed as it did so was massively destructive. If that event had occurred over a city, it would have killed as many people as an atom bomb.

As disastrous as that might have been, it was a mere ripple compared to the hit taken by one of our closest neighbours in the solar system. In July 1994, the Hubble Space Telescope observed the most spectacular explosion humans have ever directly witnessed: the destruction of Comet Shoemaker-Levy 9 as it was torn to pieces by Jupiter's gravity. At least twenty large fragments hurtled down onto the planet, their size suggesting the original comet's nucleus might have been as wide as 5km across. It took several months for the scars on Jupiter's disrupted atmosphere to fade. The rapid succession of impacts released six hundred times more energy than all our nuclear weapons arsenals combined. If even one of these cometary missiles had hit Earth, the consequences would have been far greater than the impact that wiped out the dinosaurs. No life significantly more complex than bacteria would have survived.

A comet might have been responsible for flattening 2000km² of forest in Tunguska, Siberia, in 1908.

Even so, some recent discoveries do support at least some of Hoyle's notions. The NASA space probe Stardust was launched in February 1999 to study asteroid 5535 Annefrank and collect samples from the tail of Comet Wild 2. The most important phase of its mission was completed on 15 January 2006, when its little sample capsule returned to Earth in a fiery re-entry, followed by a more gentle parachute landing in the Utah desert. Stardust had several encounters with cometary and interstellar dust during its three orbits around the sun. Analysis of captured materials revealed tar-like molecules even larger and more complex than the PAHs found in the Mars meteorite ALH84001. These weren't just hydrocarbons. Nitrogen and oxygen also featured in Stardust's samples.

There is no doubt that some surprisingly intricate organic molecules are fabricated in interstellar space by the action of starlight on dust and gas. There's no doubt that they aggregate into snowballs of cometary material. There's also no doubt that plenty of these must have fallen on the young Earth, breaking up in the atmosphere and superheating, for sure, but also releasing vapours rich in organic compounds. As these molecules encountered liquid water in clouds, or drifted to the ground, they may well have triggered some of the chemical reactions that contributed to the emergence of life.

Arguments against panspermia

Water, at least in its liquid state, is commonly regarded as essential for life. Water needs a home world to keep it warm, and a surrounding atmosphere to prevent evaporation. A dust cloud drifting in the frigid vacuum of space seems an unlikely venue for the creation of biologically advanced molecules, let alone anything so sophisticated as DNA.

Even if we allow for liquid water, the biggest problem is more subtle. How can evolution get going in a diffuse interstellar nebula or a cometary ball of ice? Why would an intricate and information-rich organic assembly, reaching towards the level of a living system, be any better at surviving in space than a duller and much more basic collection of chemicals? Once a haphazard grain–organic combination has obtained physical protection against the rigours of space, there is no particular evolutionary pressure for it to become more complicated. That, alas, is not how cruel, heartless evolution works. Simple molecules tucked safely away in the shadows can survive just as well as not-so-simple ones. Time passes, but none of them have to die.

Around 600 million years ago, the great surge of evolution known as the **Cambrian Explosion** generated countless new forms of life on Earth. It began with a relatively simple development, from organisms that hung around all day, lazily soaking up sunlight, to other organisms that evolved a taste for swallowing up their neighbours. When some of those organisms developed the trick of wriggling, swimming or crawling alongside their less agile companions, the long dance of hunter and hunted began – and with it, an accelerating arms race between predator and prey. The DNA molecule allows successful organisms to pass down winning characteristics to their descendants. Those characteristics are encoded in long and extremely specific strands of molecular information. Biology isn't defined by raw chemicals alone. It's much more a matter of the dynamic processes that the molecules' data content can unleash. This is not something that can be conjured up from simple organic compounds randomly clumping together in a dust cloud drifting through space.

Most scientists think that you need a fair-sized planet (or at least, a decent-sized moon like Jupiter's fourth largest, Europa) with liquid water before life can emerge, but panspermia certainly isn't all nonsense. Dormant bacteria could probably survive a long trip through space in the deeper recesses of meteorites. Ancient meteorite strikes might conceivably have transferred living organisms between Earth and Mars. Both worlds were heavily bombarded by meteorites until about four billion years ago while the solar system was still settling down after its formation. Cometary strikes may have brought most of the Earth's water, plus plenty of organic building blocks for life, even if it took subsequent evolutionary processes on Earth before anything came along that was actually alive.

The moon bug

The three-legged Surveyor robot probes were the first NASA spacecraft to touch down safely on the moon in the years leading up to the more famous astronaut landings. In November 1969, Apollo 12 astronauts Pete Conrad and Alan Bean walked to the Surveyor 3 spacecraft and broke off its TV camera for return to Earth in a sterile bag. NASA scientists wanted to see how Surveyor's engineering materials and electrical systems had coped after more than two and a half years of exposure to the lunar environment, but they also checked for bugs, just for the hell of it. To their surprise, a hundred or so cells of a common terrestrial bacteria, *Streptococcus mitis*, were found nestling in the camera's foam insulation. Evidently a NASA technician had coughed out a little cloud of these things while assembling the camera prior to launch. They survived the vacuum of space, the hazards of radiation exposure, and deep freeze at an average temperature of only twenty degrees above absolute zero (-273.15°C, near which almost all molecular motion ceases), with no nutrients or water to assist them.

It seems that many bacteria, in dormant form at least, can survive for long periods, perhaps thousands of years, as long as they are not exposed directly to the harsh and destructive ultraviolet radiation of the sun. Whether or not any organisms could survive tens of thousands, or even millions, of years drifting through space is not yet known. But partial panspermia, the exchange of biological material between neighbouring worlds in a solar system, is now regarded as scientifically credible.

Unexpected petri dish: the well-travelled Surveyor 3 camera kept a host of common earthly bacteria warm and cosy on the moon for more than two and a half years.

This raises an interesting problem. If we discover life on Mars containing DNA or RNA similar to the chemistry of its earthly cousins, we might merely have encountered an offshoot of life on Earth, transported across the solar system by a series of asteroidal accidents. It's even plausible that we could find fossilized mineral traces on Mars predating similar examples on Earth. Any such developments would be immensely exciting, but they would not answer two fundamental questions: how did life begin, and can it arise independently on other worlds?

What we really hope to find one day is an extraterrestrial organism that is sufficiently different in its biochemistry that it cannot by any stretch of the imagination be related to life on Earth. Such a discovery would prove that life is not a one-off fluke, and that it can arise in other solar systems too.

The cosmic haystack

Are we the only sentient and technological creatures who exist? Our quest for answers is focused mainly on the little chunk of the universe that we can investigate with our instruments: our galaxy, the Milky Way. Even in this local area, the hunt for alien life is like looking for a needle in a haystack.

Finding our way in the Milky Way

Our solar system drifts at one end of a spiral arm in a fairly typical galaxy within a universe so vast, there is no easy way for us to grasp the distances involved. Earth orbits the sun at a distance of 150 million km. This is a convenient yardstick for comparing distances in and around the solar system. The Earth–sun distance is known as an **Astronomical Unit** (AU). To get an impression of what an AU really means, think of the sun as a large sunflower and the Earth as a little blue ant about fifteen metres away. Pluto's average distance from the sun is about forty AUs. The outermost realms of the solar system, where lonely comets lurk, is somewhere around 100,000 AUs in diameter.

Beyond that vastly distant realm, even the AU loses its descriptive power as a measure of distance. Instead, we have to turn to the fastest-known entity in the universe, light, which travels one AU in eight minutes. When measuring distances in the broader universe, we need to think in terms of how far light can travel in years, decades and millennia.

The speed of light in a vacuum is 299,792km per second. In one year, an unimpeded photon (the indivisible basic unit of light energy) travels 9,460,800,000,000km, a distance termed a light year by astronomers. Besides the sun, the closest star to us is Proxima Centauri, 4.2 light years away. Imagine brilliant, glowing oranges separated from each other by the

Messier 31, also known as the Andromeda Nebula, is the nearest large galactic neighbour to the Milky Way.

distance between New York and Chicago, or between Paris and Glasgow. This gives a rough idea of the mind-numbing voids between stars that are essentially right next door to each other.

Our local galaxy, the **Milky Way**, contains at least 200 billion stars. From one side of the galaxy to the other, the distance is around 100,000 light years, and its densely populated central core of stars is about 30,000 light years deep. The light that we capture in our telescopes from stars at the galaxy's perimeter was first emitted while Neanderthal humans still roamed the Earth.

Galaxies

Shaped by the collective gravitational forces of all the stars, gas and dust clouds within them, galaxies usually adopt the form of spirals or ellipses. A spiral galaxy bulges in the centre, where most of its stars are concentrated. The rotation of the galaxy causes outer stars to be dragged along in the gravitational tide as spiral arms. The Milky Way is a spiral galaxy, and it turns a full circle once every 220 million years. Our sun, residing in one of the outer arms, has travelled around the galactic centre about twenty times since the solar system's birth five billion years ago.

A number of dwarf galaxies and diffuse clouds of stars are gravitationally linked with the Milky Way in a vague collection known as the "Local Group". The next nearest major spiral galaxy similar to our own is Andromeda, 2.5 million light years away from us. Sunlight reaches Earth in about eight minutes, already spanning in that brief time a distance greater than we can easily grasp. No human-scaled analogy can encompass a distance as vast as 2.5 million light years – and that's nearby in cosmic terms.

Galaxies are distributed unevenly in space. Some drift in splendid isolation with no close companions. Others form strange pairs, orbiting each other or, in some cases, colliding and mingling. It seems incredible, but collisions between stars are rare because of the inconceivably vast distances between them.

Many galaxies are found in groups called clusters, typically millions of light years across. A cluster may contain anything from a few dozen to several thousand galaxies. Clusters, in turn, are grouped in superclusters. On even grander scales, galaxies tend to accumulate in a tangle of filaments surrounding relatively empty regions of the universe, known as voids. One of the largest structures ever mapped, a network of galaxies known as the Great Wall, is more than 500 million light years long and 200 million wide.

The known universe – that's to say, everything in space that we can observe with our instruments – contains somewhere between 100 billion and 500 billion galaxies. There are probably more galaxies in the universe than there are stars in the Milky Way. Light from the most distant galaxies that we can observe has taken nearly thirteen billion years to reach us. That light first shone when the universe itself was still young. Because of these mind-numbing expanses of space and time, it is extremely unlikely that we will ever encounter extraterrestrial intelligences from another galaxy. The best place to look is within our own.

The Fermi Paradox: where is everybody?

Italian-born physicist **Enrico Fermi** (1901–54) escaped from Mussolini's Fascist regime and joined America's secret Manhattan nuclear bomb project in the 1940s, for which he built the world's first working nuclear reactor. He was one of many brilliant nuclear scientists of that era who feared, quite reasonably, that Nazi Germany was on the brink of developing nuclear weapons, and that this threat could not go unanswered. It came as a surprise to some of them that the eventual target for the first aggressive detonation of nuclear weapons was not Germany but Japan.

In the summer of 1950, Fermi was visiting the Los Alamos laboratory in New Mexico, where "the Bomb" was developed during World War II under the leadership of Robert J. Oppenheimer. One day, Fermi was having lunch with fellow physicists Edward Teller, Emil Konopinski and Herbert York. They chatted briefly about a newspaper cartoon mocking a recent news story. Why were New York's public litter bins mysteriously disappearing off the streets? According to the cartoon, the bins were being stolen by aliens in flying saucers. As the party strolled in the open air towards their lunch venue, they bantered for a while about little green men and the impossibility of faster-than-light travel. By the time they'd sat down, their conversation was directed towards more earthly matters until, out of the blue, Fermi asked, "Don't you wonder where everybody is?" His companions laughed at the suddenness of his interjection, but he was serious.

Fermi's inspiration: little green men stealing New York's public litter bins (by Alan Dunn, *The New Yorker*, 20 May 1950).

No one was taking minutes at this informal gathering. Many years later, Teller and Konopinski recalled the incident in slightly different ways. Both men were sure it was a summer get-together, though Teller thought it must have happened just as the war ended. Only that cartoon, published on 20 May 1950 in *The New Yorker* sets the pivotal moment in a specific year. Whatever the details, today we've inherited "Where is everybody?" as the defining version of Fermi's challenge, now known as the **Fermi Paradox**.

The word "paradox" comes from the ancient Greek for "beyond belief". In formal logic, a paradox is a statement or concept that contains conflicting ideas, as when a philosopher tells you, "my next statement will be true", then trumps it with, "my last statement was a lie." Each expression relates to the other in a seemingly sensible fashion. It's just that when we consider them together, they clash. The paradox arises because the philosopher creates a set-up that is neither true nor false. In the usual run of things, either something's true or it's not. A paradox throws such simple notions out of the window.

So why is Fermi's question described as a paradox? Essentially because, by implication, two seemingly plausible statements contradict each other. Statement one: according to our best scientific understanding, the universe should be teeming with life on other worlds. Statement two: according to our best scientific understanding, there's no sign of life apart from on this world.

"It is probable that a good many of the billions of planets in the Milky Way support intelligent forms of life. To me this conclusion is of great philosophical interest."

Otto Struve, *Astronomers in Turmoil* (1960)

Fermi and his lunch companions were interested in the specific question of aliens visiting Earth. The great twentieth-century nuclear physicists were excellent at probability theory. They were also famous for witty shorthand phrasings of deep questions, whether in mathematics or wordplay, so it's not fanciful to extrapolate a paragraph of explanation from just three words uttered by Fermi. Here is what he was really asking when he blurted out, "Where is everybody?"

The probability of "everybody"

Even if the chances of intelligent life emerging in any given star system are extremely small, there are many billions of stars in the galaxy. Clever creatures must have arisen occasionally. Among even the most advanced alien societies, it might be extremely rare to find any with an interest in costly and difficult space adventures. Of those few who do give it a try, only a tiny proportion might be sufficiently advanced for interstellar travel... and so on.

Fermi's point was that surely one civilization should be a mould-breaker, expanding into the galaxy via technologies that would make NASA's slickest probes look like school science projects. The fact that nobody has visited Earth suggests that there's no one capable of doing so. The Fermi Paradox is an apparent contradiction between the supposed probability of extraterrestrial civilizations existing, and the supposedly odd fact that we haven't had contact with any of them.

Everything depends, of course, on what we mean by "probability". No matter how smart Fermi and his friends might have been, their famous chat was little more than speculation. Ten years after the Fermi Paradox was phrased, an astronomer in Green Bank, West Virginia, decided that the hunt for aliens should be stated more formally in a scientific equation that might one day be put to the test by comparing its numbers to actual scientific observations.

The world's first alien hunter

Frank Drake (1930–) studied electronics at Cornell University, where he became fascinated by astronomy. As a teenager he'd attended a Baptist Sunday school. University trips to museum collections of Egyptian mummies and other artefacts caused him to wonder how his particular church could so easily ignore the religions of other cultures, past or present. From that thought came an insight familiar to ancient Greeks and heretical Renaissance thinkers alike. If religions were multitudinous, then perhaps the entire human world was similarly just one among many?

One visiting lecturer at Cornell particularly fired Drake's imagination. **Otto Struve** (1897–1963) spent his early years in Russia before his dramatic escape from revolutionary turmoil. He reached the US in 1921 and spent the rest of his life as a respected astronomer and administrator. Among his major contributions was the study of how stars rotate, and how their spin speed is related to planet formation. Young, hot, massive stars spin extremely fast, while modest yellow dwarves, such as our own sun, turn more sedately.

Struve showed that stars begin their lives with the spin energy, or angular momentum, retained from the vortex of dust and gas that they draw into themselves while forming. They shed some of that momentum if planetary bodies also accrete from the dust cloud, because mutual gravitational drag is exerted between those new worlds and their sun. Think of an ice rink dancer spinning rapidly on her skates, with her arms tucked close to her body. Now imagine her stretching out her arms, extending the diameter of mass distribution in the dancer-and-arms system. Her spin rate slows down. **Slow-spinning stars** most likely have a retinue of planets, Struve concluded. There are a great many slow-spinning stars in the galaxy, so planets must also be commonplace. In turn, a plethora of planets suggests no shortage of habitats where life may have taken hold.

Struve was a conservative scientist, not given to flights of fancy, yet he was one of the few eminent astronomers in the mid-twentieth century who insisted that extraterrestrial intelligence must be abundant. When young Drake heard Struve's lecture on this theme at Cornell in 1951, he was inspired to turn his purely personal interest in aliens into a formal professional quest. If Struve said that the search for alien civilisations was acceptable science, then it must be, Drake reasoned.

After three years in the US Navy, gaining first-hand experience of complex radio equipment, Drake became a radio astronomer at the newly founded National Radio Astronomy Observatory (NRAO) in Green Bank, West Virginia. The band of electromagnetic radiation accessible to optical telescopes is very narrow, so most of the universe's energies are invisible, except through instruments that can detect them. Radio astronomy is the study of distant celestial objects, such as galaxies and pulsars, that emit radio waves as part of their energetic processes. But it was a very particular kind of radio emanation – a message from an alien intelligence within our own galaxy – that Drake hoped to find.

Fortunately, in 1959 Otto Struve, an ally with obviously similar interests, was appointed director of the NRAO. Drake won permission to conduct a radio search for artificial signals, on the strict understanding that the circuitry and other apparatus he designed for the task should be used only briefly for alien hunting. The funding for his project, no matter how modest, had to be based on respectable, longer-term astronomical applications, such as studying the purely natural radio emanations from interstellar hydrogen gas clouds. Struve told Drake to be discreet. Then, with startling suddenness, he changed his mind and told Drake to tell the world.

That same year, and quite independently of Drake, two well-respected nuclear physicists also proposed a radio search for extraterrestrial intelligence. They even suggested what channel to tune into. By a lucky confluence of thinking, it turned out to be close to where Drake's equipment was supposed to tune. All of a sudden the NRAO was keen to prove, rather than hide, its alien-seeking credentials. Among the next steps in Drake's quest was his invention of a neat mathematical tool for assessing the likelihood of finding alien radio signals. Before examining the famous "Drake equation" in the next chapter, or discussing those two nuclear physicists in this one, we have to pause for a while to talk about electromagnetic radiation and that special frequency.

Light, radiation and radio waves

Why does a cloudless afternoon sky look so blue? The answer to this child-like question puts us in the right frame of mind for hunting alien signals. All **electromagnetic radiation** (EM) waves take the form of intertwined electric and magnetic fields travelling together through space while oscillating perpendicularly to each other. The shorter the wavelength, the higher the energy. Waves vary in size from very low energy radio waves the length of a city block to immensely energetic gamma rays smaller than an atom.

Einstein tells us that matter and energy are pretty much the same thing. At the gamma end of the EM spectrum, the wavelengths are so short and powerful they are indistinguishable from matter particles. Gamma rays slam into soft biological tissues as destructively as submicroscopic machine gun bullets. At the other end of the spectrum, the longest radio

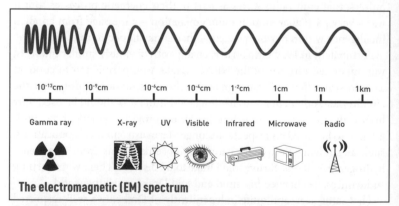

The electromagnetic (EM) spectrum

A simplified depiction of the EM spectrum, with short, ultra-high frequency wavelengths at the energetic end, and longer, lower-energy wavelengths at the radio end.

waves have so little energy they barely interact with the solid world, passing through humans and brick walls alike without leaving a trace of their presence.

All EM radiation is light, even though we can only see a limited range of it through our eyes. The fundamental carrier of light is the photon, a particle that has no mass but can convey energy. Every photon begins its journey when an electron (one of the tiny, negatively charged specks of matter that orbits the nucleus of an atom) is boosted into a higher, more energetic level by being heated, or by absorbing another photon from somewhere else. The electron eventually falls back to its original state, releasing its excess energy in the form of yet another photon. The more energy absorbed by an electron in the first place, the higher the energy of the photon it releases. Photons behave like waves or particles, depending on what kind of instrument observes them. This **wave–particle duality** is one of the central mysteries of science, and it's all very interesting, but we'd better get back to the business of the blue sky and its relevance to alien signals.

Sunlight containing a wide range of EM frequencies reaches the upper layers of the atmosphere, but then molecules of gas, plus suspended water vapour, dust and ice crystals, scatter the wavelengths in different ways. The nasty, destructive stuff, including the biologically hazardous shortwave ultraviolet energy, is absorbed before it reaches the ground, a blessing that has been crucial to the emergence of life on Earth. The shortish optical wavelengths of blue light arriving from the sun do penetrate deeper, but are bounced around and scattered by atmospheric water vapour, causing the sky to appear blue in daylight hours. When the sun sets, and its light makes its longest and most oblique journey through the air, vapour and dust, almost all of the blue light is scattered, and only the long, languorous wavelengths of red light reach our eyes. That's why sunrises and sunsets look so red.

What this tells us is that an alien civilization wishing to beam a signal deep into space might favour longer wavelengths of EM that can slip past interstellar dust clouds and other interfering clutter. Space-based telescopes scan distant celestial targets at infrared frequencies (the EM energy given off by warm or hot objects) precisely because those long wavelengths slip past most of the dusty and gaseous obstructions in the voids of space. In terms of signalling, the most realistic frequencies are even lower, among the radio realms of the spectrum. One frequency in particular was identified in 1959 by that pair of eminent physicists, even as Drake made ready with his project at the NRAO.

Naturally regular signals

Just because a cosmic signal is neat and regular doesn't necessarily mean that it's artificial. In 1967, Jocelyn Bell, a young Cambridge University radio astronomer, detected a regular and fast-cycling radio pulse coming from a pinpoint source in the sky, as if from an alien transmitter. She and her supervisor, Anthony Hewish, scribbled a light-hearted note: "LGM!" (Little Green Men!)

One of the intriguing aspects of the source was its omnidirectional nature. It was tempting to imagine that it was a giant broadcast beacon or navigational aid of some kind, or perhaps an alien greeting that took into account the difficulty of aiming messages at specific targets in the vastness of the cosmos. It was as if an alien intelligence had set up an illuminated signpost in the night, careless of the identities of those who might read it, but wishing to attract attention just the same.

Bell never really believed she'd discovered aliens, but the signals were unusually regular, just as if they'd been generated by a machine. Day after day, they pulsed at exact intervals of 1.3373 seconds. The signals actually came from a pulsar, a small, extremely dense star that spins around its axis in just a few seconds, sending out intense beams of radiation like a lighthouse gone mad. A powerful gravity field holds the star together, countering the centrifugal force of its breakneck rotation. Since 1967, over three hundred pulsars with similar exactly spaced signals have been discovered within the Milky Way, each with its own time signature.

The search for interstellar communications

Italian physicist Giuseppe Cocconi (1914–2008) was a senior figure at Europe's CERN laboratory (the European Organization for Nuclear Research) throughout the 1960s. His suggestion that we might be able to talk with aliens, or at least listen in on their chatter, never hindered his career as a respected scientist. None of his colleagues mocked him, or questioned his fitness to work at the frontiers of subatomic physics. If Cocconi had aired his views on extraterrestrials in, say, 1950, he would have been derided, but by 1955 the scientific community was more receptive to this subject. Space travel was no longer a matter for science-fiction fans alone. It was quite clear that satellites, and even human-carrying capsules, were imminent. Space promoters such as Arthur C. Clarke even held out the prospect of putting people on the moon by the end of the twentieth century. Events moved faster than even Clarke had dared hope. The moon was within our reach by the end of the 1960s.

In the febrile yet undeniably thrilling era of technological advance into space, extraterrestrial intelligence could be discussed in earnest rather than merely as a fictional conceit. Just before Cocconi began his career with CERN, he met Philip Morrison (1915–2005), a physicist at Cornell University and a veteran of the Manhattan Project. In September 1959, these two sober and respectable scientists wrote a ground-breaking paper for *Nature*, one of the handful of journals (along with the similar US publication *Science*) that are regarded as international benchmarks for the announcement of new scientific ideas. The paper was called "Searching for Interstellar Communications".

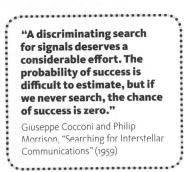

"A discriminating search for signals deserves a considerable effort. The probability of success is difficult to estimate, but if we never search, the chance of success is zero."

Giuseppe Cocconi and Philip Morrison, "Searching for Interstellar Communications" (1959)

Their opening paragraph assumed that among the many billions of suns in the galaxy similar to our own sun, it was not unreasonable to expect that a few might harbour planets like our own, where intelligent, technological societies could exist. It was a brief reformulation of the Mediocrity Principle (see p.27). Taking it as read that such civilizations might be out there, Cocconi and Morrison moved on to the real subject of their paper: the best way to detect the existence of such civilizations. "Interstellar communication across the [galaxy] without dispersion in direction and flight time is practical, so far as we know, only with electromagnetic waves", they wrote. This was the first formal scientific proposal to search for aliens by tuning a radio to the right station.

What's the frequency?

Anyone who has fiddled with the dial of an old-fashioned radio, or even channel-hopped restlessly through the menu of a digital one, will spot the major problem. On what specific channel might the aliens be broadcasting? Cocconi and Morrison assumed that smart aliens would compensate for background radio noise in their own neighbourhood, so that any signals deliberately pulsed into space would not be confused with the random background roar of their own galactic neighbourhood.

Local interference is just the first hurdle to overcome. Throughout the galaxy, billions of stars, black holes, white dwarfs, planets, moons, dust clouds and even frozen comets all emit radiation. Everything, everywhere, is slightly warm. Infrared energy is, essentially, heat radiated from anything that isn't absolutely cold. A cube of ice fresh out of the freezer may feel cold to our human touch, but we have to plunge down another 273°C in temperature before we reach the theoretical point when an object is truly cold and emits no radiation at all. Nothing in nature ever quite reaches absolute zero. The incessant restlessness of atoms forbids it. Traces of infrared warmth saturate the cosmos, and at even lower energies, the electromagnetic hiss of radio waves creates a drizzle of multi-frequency white noise coming from all regions of the sky at once.

Amid that random cacophony, from loud noises to faint static crackles, how are we to find the needle of an alien transmission frequency among the haystack of electromagnetic possibilities? Cocconi and Morrison suggested "a unique, objective standard of frequency that must be known to every observer in the universe". They defined that frequency as 1420 megahertz, with a wavelength of 21cm. This is exactly the kind of very low energy signal that can travel vast distances across the galaxy without being absorbed by dust and gas clouds. But that's not the only reason why it's special. It is called the hydrogen radio emission line, or simply the **hydrogen line**.

The hydrogen frequency

Hydrogen is the fundamental building block of the cosmos. What we think of as the empty space between stars is suffused with countless trillions of tons of hydrogen in the form of nebulous clouds. In terms of the human-scaled world, these ghostly entities are vanishingly thin. An astronaut waving a gloved hand through the vacuum of space might unwittingly encounter a few dozen hydrogen atoms, but from her point of view, there would seem to be nothing in her hand. At the grand cosmological scale, this perspective is reversed. If the galaxy were to ponder its own composition, it would regard itself as a swirling mass of hydrogen. Stars are little more than glowing fleas in the hydrogen fur of creation, and the occasional astronauts count as the galactic equivalent of "nothing".

The single electron buzzing frenetically around the perimeter of a hydrogen atom shifts restlessly between energy states, creating a weak but detectable electromagnetic presence. In the very low-energy regions of the radio spectrum, where the galaxy is at its quietest and most serene, the faint whisper of the hydrogen clouds can be heard. If an unusually strong source of radio emissions was discovered at that frequency, coming from a highly localized source rather than a vague mass of hydrogen gas, it would be very hard to conceive of any natural object that might account for it.

The redshift

In 1929, American astronomer Edwin Hubble (1889–1953) demonstrated that the universe was inconceivably vaster than anyone had imagined. Using the newly built 2.5m-diameter Hooker Telescope at the Mount Wilson Observatory, California, he showed that all those small dots of light in the night sky aren't necessarily just stars and planets. Many of them are galaxies, as vast as our own Milky Way, and immensely far away. Hubble also proved that all galaxies are flying apart from each other at colossal speed.

The further away a galaxy is, the faster it recedes from us. The light from distant galaxies stretches into longer wavelengths at the red end of the spectrum, just as sound waves from a train's whistle are stretched into lower frequencies as the train hurtles past us and speeds off into the distance. This phenomenon is called the redshift, and it affects every facet of astronomical science, including the hunt for alien signals in radio frequencies.

Edwin Hubble preparing to take photos of the sky from the small (48-inch) Schmidt Telescope in California.

From our point of view, the hydrogen frequency shifts depending on how far away any particular emission source happens to be from Earth. The constant expansion of the universe stretches ("redshifts") any radiation into longer wavelengths as the distance between the source and the observer increases. That said, redshift only becomes markedly significant when we are observing extremely distant objects. We assume that intelligent aliens would do the sensible thing and broadcast at the neutral frequency of the non-redshifted ("neutral") hydrogen sources near them. That frequency would be just the same as the one generated by non-redshifted hydrogen sources close to Earth. This is why the neutral hydrogen line makes a reliable yardstick for all observers, whether earthly or not. If an extraterrestrial civilization is using radio technology to probe the depths of space, they will have worked it out.

Listening
for a signal

In the summer of 1960, the first formal radio search of two nearby star systems was conducted in a bid to find signs of sentient alien life. A new astronomical discipline was born that year: the Search for Extraterrestrial Intelligence, or SETI.

Project Ozma

In August 1960, Frank Drake and his NRAO allies (see Chapter 7) conducted Project Ozma from Green Bank, West Virginia. According to Drake, the project was named "for the queen of the imaginary Land of Oz, a place very far away, difficult to reach and populated by exotic beings". Ozma took the form of a two-week radio dish observation of Tau Ceti and Epsilon Eridani, two relatively close sun-like stars (respectively, 11.8 and 10.5 light years from our solar system). On the first day, shortly after turning the **Green Bank telescope** towards the second star, they heard a signal that seemed to come from Epsilon Eridani. According to Drake, "We heard bursts of noise coming out of the loudspeaker eight times a second, and the chart recorder was banging against its pin eight times a second. We all looked at each other, wide-eyed. Could it be this easy?"

Controlling his excitement, and being very careful not to make any rash announcements, Drake immediately set his mind to an important problem: how could they avoid false trails and be reasonably sure that any exciting signal they picked up was artificial and not of human origin? The first check was to swivel the dish off-beam from Epsilon Eridani. If the signal disappeared as soon as the dish looked in another direction, then this would suggest that the signal did indeed emanate from the vicinity of that star. Right on cue, the Epsilon Eridani signal was lost as soon as Green Bank's dish averted its gaze. Unfortunately, when Drake steered the

Did the Ozma signal come from a Jupiter-like giant in the Epsilon Eridani system (artist's impression) or a smaller companion world?

dish back on track, the signal was gone, so he had no proof that waving the dish back and forth had made any difference. Perhaps, coincidentally, the signal just stopped of its own accord?

By the time he heard the signal again, Drake had taken the further precaution of aiming another, simpler antenna at a wide region of the sky. It picked up the pulses too, proving that it must be interference from something terrestrial, rather than the longed-for sign of alien life. Drake eventually discovered that the mysterious signals were probably from a military aircraft testing radio jamming equipment overhead, "because the people involved thought it would do the least harm in that remote region of West Virginia".

The project was a success of sorts – real science in action. Eliminating false data is an essential step towards finding what you're after. Drake and his modest Ozma team devoted two hundred hours of observation to their two targets. They scanned around the hydrogen line frequency of 1420MHz, making allowances for redshift drifts that might result from the relative motions of the Earth and any supposed signal sources, just as Morrison and Cocconi had suggested (see p.127). While Ozma didn't find a signal from an extraterrestrial civilization, it did become the model for future radio-based SETI projects.

Following on from this flurry of activity, the US National Academy of Sciences suggested that Drake arrange a conference of scientists interested in furthering the search. This was at a time when newly fledged space agency NASA was making plans for robotic Mars missions. Many scientists expected to find lichen or other simple life forms on the Red Planet, and now seemed a good time to establish useful procedures for experimentation, whether on Mars or further afield.

Drake's eleven-strong delegate list included Otto Struve, Philip Morrison, Giuseppe Cocconi, and a young Carl Sagan, whose expertise in planetary atmospheres made him a useful member of this strange new gang. Drake also invited Bernard Oliver, a research director from Hewlett-Packard, and Melvin Calvin, whose experiments in biosynthesis were similar to Miller and Urey's (see p.38), except he'd used radiation rather than electric sparks to set things in motion. Also on hand was dolphin communications researcher John Lilly, whose work seemed to offer a relevant earthly parallel with the challenge of talking to an alien intelligence. The group nicknamed itself the "Order of the Dolphin". The three-day meeting, held at Green Bank, West Virginia, in November 1961, established the groundwork for the exploratory science now known as SETI, **the Search for Extraterrestrial Intelligence**.

The Drake equation

Ahead of the conference, Drake decided that the chances of finding alien radio signals had to be phrased as a statistical probability. If he could put the problem to sceptical fellow scientists in a formal mathematical language, rather than at the level of mere philosophical speculation, then all the variables could be assessed one by one. He devised a famous equation that still bears his name. Here's what it looks like (don't be daunted – it's actually quite simple):

$$N = R^* \; fP \; nE \; fL \; fI \; fC \; L$$

R^* the rate of formation of stars suitable for the development of intelligent life

fP the fraction of those stars with planetary systems

nE the number of planets, per solar system, with an environment suitable for life

fL the fraction of suitable planets on which life actually appears

fI   the fraction of life-bearing planets on which intelligent life emerges

fC   the fraction of civilizations that develop technologies that we can detect via radio

L   the length of time that such civilizations release detectable signals into space

N, the magic figure we're after, is the number of civilizations whose radio pulses we might detect.

At each point, the numbers must be shaved down ruthlessly. (R*) can't include the many billions of stars that are improbable hosts for life-bearing planets because they are too hot, too unstable or giddily swirling around in binary star systems. And within (P) – solar systems that are remotely like our own – Jupiter-sized gas giants, sun-blasted inner planets and icy outlanders are discounted as likely habitats for life. Even if worlds exist that are quite like Earth, allowances have to be made for the fact that life might not necessarily arise on them (L). And if creatures do evolve, we have to be pessimistic about the likelihood of any becoming intelligent (I). Even if some aliens do stare up at their skies in fascination, we have to wonder if advanced technology is within their grasp, or on their wish list (C); they may have no interest in radio dishes. And even if – and it's a big if – they are technological, and do fire "hello" messages into space, the final question (L) asks when, and for how long? This year? Next year? Ten thousand years ago? Did they send a short, brief burst of communication that we've already missed, or are they even now transmitting from a beacon built to last for millennia?

The universe is vast in time as well as space. Let's conjure up what Einstein would have called a *Gedankenexperiment*, or "thought experiment". Suppose that on a beautiful, Earth-like planet just twenty light years away, a technological civilization evolved at almost exactly the same rate as ours. Their classical scholars mastered astronomy at the same time as ours, give or take a generation. A hundred years ago, plus or minus a decade or two, their engineers invented radio. Half a century ago, give or take a few years, they beamed a message into space, hoping that someone like us would hear it.

Give or take... plus or minus... approximately... On the cosmic scale of events, measured in billions of years, this imaginary civilization and ours have emerged, against all odds, with fantastic synchronicity. On the puny human scale of time, the slightest chronological misalignment between their technology and ours makes all the difference between setting up a premium-rate interstellar chat line, or passing each other like ships in the cosmic night, never knowing how tantalizingly close we came to hooking up. They might have been sending out artificially modulated gravity waves or greetings encoded in quantum entanglement, while we were still fooling around with crystal set radios. We might try to beam terabytes of data in laser pulses towards their star system, just as they're mastering the equivalent of Morse code.

Finally, the time it takes for signals of any kind to cross interstellar space adds a complicating factor of anywhere between four years (the distance in light years to our nearest neighbouring star, Proxima Centauri) and many thousands, depending on the location of individual stars within the galaxy. And that's just the local stars. Beyond that, the distances become so vast that the technological challenges of detecting signals – let alone exchanging messages back and forth – stretch beyond our reach. The galaxy may be teeming with intelligent, technological civilizations, all eager to make contact with other worlds, but there's no guarantee that we'll ever be able to communicate with them.

Soviet SETI

During the Communist era, the spiritual instincts of the Russian people were redirected away from the rival power structure of the Russian Orthodox Church and towards a more materialist way of thinking. Factories, power stations and space rockets were the new cathedrals. Notions of interplanetary travel and the possibility of life on other worlds slotted surprisingly well into Soviet philosophy. Communism would reach for the stars, and would one day find advanced alien societies built on similarly socialist lines, thereby proving that socialism was the right course.

The first Soviet SETI conferences were held at the Byurakan Astrophysical Observatory in Armenia in 1964 and 1971. Foreign delegates included Sagan, Drake and Morrison. The Soviet Academy of Sciences sanctioned occasional SETI search programmes under the leadership of respected radio astronomers, including Aleksandr Leonidovich Zaitsev, V.S. Troitsky, Iosif Shklovskii and Nikolai Kardashev. Drake acknowledged at the time, "the great respect paid to SETI and to science in general in the Soviet Union", but worried that "there was far too little mutual contact, mutual criticism and peer review in Soviet science. The papers ranged from excellent to preposterous." Nevertheless, some intriguing philosophical ideas emerged from these tentative contacts between East and West (see the Kardashev Scale, p.183).

After Ozma

A decade after Drake's pioneering first search, NRAO astronomer G.L. Verschuur led a second SETI attempt at Green Bank, targeting nine nearby stars, and once again monitoring in and around the hydrogen line frequency, even though he wasn't convinced it was the right channel. He wrote at the time, "Any detection of a signal from another civilization will most likely be an accidental one in the sense that we will pick up signals not meant for us. For this reason it is unlikely that [the hydrogen line] is the wavelength at which to search." He had a point. But a bigger problem

The "Wow!" signal

On the night of 15 August 1977, at Ohio State University's Big Ear radio observatory, volunteer observer Jerry Ehman detected a signal apparently coming from somewhere far out in space. It had the characteristics one might expect from an extremely specific "narrow bandwidth" deliberate transmission. In those pre-plasma screen days, radio data was recorded as a grid of numbers on rolls of paper print-out. Ehman circled the mysterious pulse and wrote "Wow!" on the margins of the paper. Ever since, it's been called the "Wow!" signal. Unfortunately, it never arose again, so no one is sure what generated it. And the frequency? As near as makes no difference, it was on the neutral hydrogen line.

The "Wow!" signal is SETI's greatest enigma, though no one has proved it was of alien origin. In 1994, Ehman cautiously told Cleveland newspaper *The Plain Dealer*, "We should have seen it again when we looked for it fifty times. Something suggests it was an Earth-sourced signal that simply got reflected off a piece of space debris." The problem is that no one can account for a terrestrial signal at that particular frequency bouncing in such a way, especially given the fact that all terrestrial broadcasters, both military and civilian, carefully avoid transmitting on the hydrogen line, precisely because it's so important for astronomers and space scientists. In a very technical 1997 reanalysis of the data, "The Big Ear Wow! Signal: What We Know and Don't Know About It After 20 Years", Ehman concluded, "all of the possibilities of a terrestrial origin have been either ruled out or seem improbable."

may have been that his search totalled just thirteen hours over two years, whereas Drake's team had devoted a hundred and fifty hours over three months. The next search programme, **Ozma II**, ran intermittently from 1972 until 1976. Snatching occasional spare hours from more conventional workloads, NRAO observers surveyed 674 stars for a total of 500 hours. Early SETI efforts always had to compete for telescope time with mainstream astronomical surveys.

Big Ear: the first long-term SETI project

The longest-running and most famous search was conducted over two decades, and more or less continuously, by the huge Big Ear radio telescope at Ohio State University. The instrument was a fairly simple flat aluminium surface the size of three football fields, with a giant reflector at each end, one flat and one parabolic. Big Ear's first decade was spent surveying natural radio sources (such as distant galaxies with energetic cores) until Congress cut the funding in 1972. Then the telescope was devoted to SETI, surviving on slender NASA grants until the mid-1990s, when it was dismantled to make room for a golf course and housing estate.

NASA's listening programmes at Ames and JPL

In 1971, NASA's Ames Research Center in California initiated a study of SETI strategies and the likelihood of contacting an alien civilization. That summer they held a seminar co-sponsored by Stanford University and championed by Bernard M. Oliver, the Hewlett-Packard executive who'd attended Drake's 1961 Green Bank conference. The result was **Project Cyclops**, an unrealistically grand proposal for a field of a thousand large radio dishes occupying an area about 10 kilometres in diameter. The multi-billion dollar project was far beyond anything NASA could afford, though the agency continued to sponsor SETI workshops, from which two potential search strategies emerged. Ames favoured the specific targeting of relatively nearby sun-like stars, while NASA's Jet Propulsion Laboratory (JPL) in Pasadena, California, argued that it was pointless to speculate where alien

> "Since the possibility of an extraterrestrial origin has not been able to be ruled out, I must conclude that an ETI (ExtraTerrestrial Intelligence) might have sent the signal that we received as the Wow! source."
>
> Jerry Ehman, astronomer, 1997

civilizations might be found. Better instead to search the entire sky for a signal on the widest band of frequencies possible.

By the end of the 1970s, NASA had a semi-official SETI plan taking in both of those strategies. The **Microwave Observing Program** (MOP) was established to conduct the search, following a period of research and development. Opportunistic politicians seized on the chance to savage NASA for wasting taxpayers' money, and the MOP quickly attracted critics, but Carl Sagan assembled a petition of support, signed by many leading scientists, including seven Nobel laureates. The publicity kept the development of NASA's SETI programmes on track for another decade.

In 1992, after $60 million of investment over the previous two decades, the twin NASA searches finally began. Ames scanned more than 800 specifically targeted stars from the 305m (1000ft) radio telescope in Arecibo, Puerto Rico (the largest radio dish in the world). JPL, meanwhile, began mapping the skies using the 34m dish at the Deep Space Communications Complex at Goldstone in the Mojave Desert. The searches were also given a new NASA designation, the **High Resolution Microwave Survey** (HRMS).

Both strategies exploited advances in radio technology and computing to such an extent that Morrison and Cocconi's hydrogen line became just one in a vast sprawl of frequencies SETI snoopers could consider. As the digital era got under way, new processing equipment also became available; in particular, a portable analyser – Suitcase SETI – developed by Harvard physicist Paul Horowitz, which could scan 130,000 channels simultaneously. Just a year after they'd begun, both NASA searches fell victim to a further congressional purge, and all the agency's SETI programmes were terminated in 1993. One impatient senator complained, "Millions have been spent and we have yet to bag a single little green fellow."

Phoenix and Serendip: private sector SETI

Fortunately, NASA was never the only source of cash for SETI projects. **The Planetary Society**, founded by Carl Sagan and like-minded colleagues in 1980 to promote public awareness of space and win support for future space exploration, began campaigning on SETI's behalf. As a consequence, some of the equipment used in the NASA Ames programme was passed onto the privately funded SETI Institute, which then launched its own targeted search, appropriately named Project Phoenix after a mythical winged beast reborn from the flames of destruction. Phoenix began in February 1995, using the Parkes 210ft radio telescope in New South Wales, Australia, the largest of its kind in the southern hemisphere. The opening phase checked two hundred stars over sixteen weeks. Phoenix then reverted to its home base at Green Bank from September 1996 until April 1998, sharing telescope time roughly half-and-half with other radio astronomy projects.

The Planetary Society still supports **Project SERENDIP** (Search for Extraterrestrial Radio Emissions from Nearby Developed Intelligent Populations), a radio all-sky survey operated by the University of California, Berkeley, since 1979. (Serendip also happens to be the ancient name for Sri Lanka, from which we derive the word serendipity, or "lucky discovery".) The SERENDIP receiver sits at the focus of the huge Arecibo Observatory dish, clustered alongside other people's instruments, and listens to 168 million channels within a band of frequencies centred on the hydrogen line. The SETI team have little control over where their instrument is steered. Arecibo is built into a natural crater; the Earth itself tells the dish where to point. SETI scientists simply listen to whichever patch of sky the dish happens to be pointed towards, while other astronomical researchers get on with their work unimpeded.

As for NASA, it hasn't lost interest in SETI, but has redirected its efforts in a subtle way that sets clearly achievable scientific goals, which can be better sold to politicians, and with only minimal reference to the search for intelligent alien life. The Origins programme is a catch all label for any science conducted by NASA's various space telescopes and planetary probes that has some broad link with the question of life's origins on Earth, or its possible existence in deep space. A wide range of astrobiology programmes manage to survive as NASA adjusts to the economic problems facing the US federal system as a whole.

The SETI Institute

One organization, in particular, coordinates the many and diffuse SETI projects around the world, and is involved with everything from conferences and theoretical studies to long-term observations and permanent antennae complexes. This is the SETI Institute, founded in 1984 by Frank Drake, in conjunction with Californian aerospace engineer Thomas Pierson and astronomer **Jill Tarter**. The SETI Institute is based in a modest suite of offices in Mountain View, California. Its mission is "to explore, understand and explain the origin, nature and prevalence of life in the universe". It's a private, non-profit organization, reliant largely on financial aid from its many sponsors. Jill Tarter directs the Center for SETI

Research, while **David Morrison**, a senior astrobiologist from NASA Ames, leads the Institute's Center for the Study of Life in the Universe. Between them, these two linked entities employ more than a hundred scientists, educators and support staff. Frank Drake, currently emeritus professor of astronomy and astrophysics at the University of California, Santa Cruz, keeps his chair warm on the SETI Institute's board of trustees.

After completing a degree in engineering physics at Cornell University, Tarter obtained a master's in astronomy from the University of California, Berkeley, where she coined the term "brown dwarf" to describe under-developed stars that – so far as we can tell – don't gain sufficient mass to reach ignition point. She was the lead scientist for NASA's SETI projects until federal funding was cancelled in 1993, and the inspiration for the main character (played by Jodie Foster) in the movie *Contact* (1997). In 2004, *Time* magazine named her one of the hundred most influential people in the world, because of her tireless efforts on behalf of US science teaching.

It's never fair to highlight just two or three individuals out of an organization comprising more than a hundred dedicated and talented

SETI astronomer Jill Tarter perched atop the huge Arecibo radio telescope in Puerto Rico.

How you can take part

Modern SETI systems listen in on millions of channels. It's unlikely that any human observer could chance upon something significant in all that noise. Supercomputers are required to analyse the masses of data received. Launched in 1999, the popular **SETI@home** project, an offshoot of Project SERENDIP, was an attempt to end SETI's dependence on rare and expensive digital processing facilities. Packets of unscrambled raw radio data are sent to participating PC and Mac users around the world. Over three million users are registered, and a competitive aspect encourages continued participation. Whenever home computers aren't being used for urgent work or equally essential gaming, the spare capacity is dedicated to sifting through packets of SETI data, automatically hunting for potentially significant patterns. Individually, each computer is just a small contributor to SETI. Collectively, they pack an immense computational punch.

A recent and related scheme by the SETI Institute, **setiQuest**, is an open-source software initiative that enlists programmers, amateur developers and SETI enthusiasts to help improve SETI search algorithms and software, and to get involved using smartphone apps. Jill Tarter and her colleagues say that the success of the open-source Linux operating system inspired them to try a similar approach.

setiathome.berkeley.edu
setiquest.org

staff, but few people would begrudge a name-check for **Seth Shostak**, SETI's senior astronomer and, like Tarter, an invaluable promoter of space and science education. Before his involvement with SETI, Shostak spent time in the Netherlands studying the motions of distant galaxies. This work contributed to our understanding of the universe at its grandest scales. Drake, Tarter and Shostak are tireless representatives for the work of all their SETI colleagues. Although not involved in its day-to-day affairs, several Nobel Prize winners cast a fond eye over the SETI Institute, including Charles Townes, inventor of the laser, and Baruch Blumberg, discoverer of the hepatitis B vaccine.

Interferometry: one big instrument from many small ones

In addition to her role at SETI, Tarter serves on the management board for the private, non-profit **Allen Telescope Array (ATA)**, a field of small dishes with the express purpose of conducting astronomical research and simultaneously searching for signs of extraterrestrial intelligence. The array is a joint effort between the SETI Institute and the Radio Astronomy Laboratory at the University of California, Berkeley, and is named in honour of benefactor Paul Allen. Remember him? He was the other guy responsible for the rise of Microsoft, but he retired in 1983 to pursue other interests, carefully taking his shares with him. A multi-billionaire, Allen has supported museums, movie studios, sports teams, private space initiatives – including the prototype for the Virgin Galactic suborbital rocket plane – and countless other projects that take his fancy.

Construction of the ATA began in 2004 at the Hat Creek Radio Observatory in California, about 480km northeast of San Francisco. The complete array was expected to comprise more than 300 dishes, each 6m wide, and to occupy a square kilometre of ground. The first 42 dishes began collecting data in October 2007, but funding shortfalls at Hat Creek led to the entire array being put into hibernation in April 2011. Allen's foundation sank $30 million into the ATA. Despite this generosity, the worldwide recession has left its gloomy mark on the project, and we'll have to hope for better times ahead. In 2011, the ATA was kept alive by individual donations amounting to more than $200,000. Jodie Foster's cash contribution and passionate letter in support of SETI was a welcome touch.

The idea of the array is that large numbers of small dishes can add up to the collecting power of one enormous dish, but at less expense: a concept inherited from NASA's Cyclops proposal, only on a more realistic scale and budget. Complications can arise when you wire up all the dishes so that their individual radio detections merge together into one tremendously magnified signal. This difficult technique is known as **interferometry**.

Astronomers in the twentieth century built ever-larger telescopes and radio dishes in an attempt to gather as much electromagnetic energy as possible from faint sources in the sky. Now the tendency is to try to combine many small instruments via interferometry. Imagine two stones thrown into a pond at the same time. Each impact sends waves of disturbance towards the other. Inevitably the waves meet and interfere. Where peaks overlap with peaks, the colliding waves reinforce each other and each newly combined peak becomes taller and more energetic. Conversely, where a peak meets a trough, the combined wave form cancels out and the water becomes flat.

One big instrument from many small ones: the Very Large Array in New Mexico is a prime example of interferometry (though it's never been used to hunt for extraterrestrials).

Now think of dozens of radio dishes, or optical telescope mirrors, collecting waves from space. Complex electronics gather and superimpose each dish's harvest, so that constructive interference boosts the overall signal strength. By combining signals from many small collectors spread out over a large area, interferometry creates one large, composite image resembling what a single extremely large instrument could deliver. It's just that the image has to be constructed from all those shimmering multiple inputs.

Interferometry can also be used to suppress certain aspects of an image while strengthening others. **Exoplanets** (planets outside our solar system) are almost impossible to see in an optical telescope because the bright light from their parent stars swamps the instrument. Seen from immense distances, stars and their planets are, essentially, the same pinprick of light in the sky. An interferometric telescope exploits the fact that stars don't emit much infrared heat. A star's energy is so intense, infrared doesn't get much of a look-in. But a planet is a much cooler object, warmed by its parent star, so it will give off an infrared signature.

The same interferometry process that piles waves together and boosts them can be used to create the opposite effect, destructive interference. By boosting the infrared light from an extrasolar system and suppressing the glare of the star at other frequencies, astronomers can occasionally distinguish exoplanets from their parent stars. It's also sometimes possible to use a black mask, called a coronograph, to blot out the starlight in the field of view, essentially creating an artificial lunar eclipse and applying it to an alien sun. In 2008, Greek-American astronomer Paul Kalas used these techniques in alliance with the Hubble Space Telescope, capturing the first visible light image of a world beyond our solar system: **planet Fomalhaut b** orbiting the star Fomalhaut, 25 light years away.

A universal language?

If we do detect an intelligent signal from the deepest realms of space, what kind of information might we be able to glean from it, given the fact that it would have been generated by aliens who cannot possibly share any of our spoken languages? Most SETI scientists think they know.

We humans count in base ten because we have ten fingers, and this is our convenient method for parcelling up large numbers into something we can, quite literally, handle. Computers count in base two, so that electronic calculations can be made with simple on–off pulses. Whatever system you use for totting up your totals and keeping tabs, the fundamental truths of **mathematics** remain the same. Imagine three apples on a plate. Now try to give two people an equal share of the apples without cutting any of the apples into slices. No matter if you're a two-bit computer intelligence on this world or a seventeen-fingered philosopher from another, your counting procedures will tell you the same thing: it can't be done.

The human brain is attuned to mathematical orderliness for good evolutionary reasons. A round apple is likely to be better than a bumpy one, because bumps may signify disease or insect blight. A fallen plum, bruised and covered in dirt, is less enticing than a smooth clean one still hanging from its branch. A fish with scabby scales and one eye doesn't make an appealing catch. Well-balanced features even influence our selection of partners, because symmetrical bodies tend to signify healthy genes. These are simplistic examples, but evolution has driven us to take pleasure in beauty, balance, harmony and other aesthetic qualities because they usually imply something beneficial for us as animals. So, what happens when our instinctive desire for beauty becomes intellectually abstract rather than purely survivalist? Then we encounter scientists saying that an elegant piece of mathematics "feels" just right and "fits" neatly with their observations of the real world.

Time and again, mathematical abstractions invented by the human mind turn out to have some connection with physical reality. A famous example is a special type of carbon molecule. In 1949, US architect Buckminster Fuller (1895–1983) devised his famous geodesic dome technique to enclose large communal spaces, such as sports halls and exhibition arenas, without the need for internal walls. The domes are made from adjacent flat triangular panels. Sometimes the panels are accumulated into polygons and hexagons, but the secret of a dome's strength is that the triangular building blocks spread the load across a structure that approximates, as closely as possible, a portion of a sphere. In theory, there's no limit to the size of a dome, because it can't collapse so long as the triangular sections are rigid. Future lunar cities and Mars colonies could be encased in geodesic domes. Fuller discovered that a supposedly abstract concept lent itself to real-life applications. He never expected anything in deep space to have found the same principle.

Thirty years later, however, astronomer-chemist Harry Kroto at the University of Sussex looked at the microwave spectra of carbon-rich stars, and discovered a new form of carbon, although he wasn't sure what kind of molecular shape it took. He soon found out, by blasting samples of carbon with powerful lasers. When carbon is heated into a gas, some of the atoms gather into a stable ball-shaped molecule of exactly sixty atoms, shaped just like one of Fuller's geodesic domes. Kroto and his co-researchers Bob Curl and Rick Smalley of Rice University named this new form of carbon Buckminster-fullerine, or the **Buckyball**. The three men shared a Nobel Prize for their troubles. In 2010, NASA astronomers found exactly the same

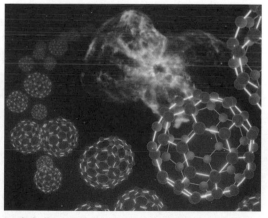

Buckyballs: artist's impression of the football-shaped carbon molecules discovered by NASA's Spitzer Space Telescope in 2010.

spectral signature in a supernova remnant. Kroto was delighted: "This provides convincing evidence that the Buckyball has – as I long suspected – existed since time immemorial in the dark recesses of our galaxy."

No wonder so many scientists think that mathematics must imply truth. In May 1963, Nobel laureate Paul Dirac, the discoverer of anti-matter, wrote in *Scientific American*:

> "It is more important to have beauty in one's equations than to have them fit experiment. If there is not complete agreement, the discrepancy will get cleared up with further development of the theory."

The mathematician's mathematician Godfrey Hardy said in 1941, "Beauty is the first test. There is no permanent place in this world for ugly mathematics." Albert Einstein was never lost for words on the subject of his fascination for pretty equations: "It is possible to know when you are right way ahead of checking all the consequences. You can recognize truth by its beauty and simplicity." That great man also pointed out that it isn't the laws of nature that are a mystery, so much as the fact that we can discern any of those laws using mathematics.

British physicist Stephen Hawking once asked, "What is it that breathes fire into the equations and makes a universe for them to describe?" This connection between imaginative constructs of the human mind and phenomena in the real world is deeply puzzling. Not everyone trusts it, but the vast majority of scientists do. Accordingly, they assume that any alien civilization with technological capabilities will have encountered the same mathematics.

A few months after Morrison and Cocconi suggested monitoring the hydrogen line for alien signals, US physicist Edward Purcell, the Nobel Prize-winning discoverer of nuclear magnetic resonance, wrote

Neptune's numbers

Here's another example of mathematics' uncanny relationship with the cosmos. Neptune, a similar world to Uranus, orbits the sun once every 165 years. Unlike all the other planets in our solar system, it is invisible to the naked eye because of its extreme distance from Earth. It was first seen through a telescope in 1846, by German astronomer Johann Gottfried Galle. Thanks to the calculations of French mathematician Urbain Le Verrier, astronomers knew exactly where to find the planet, even though they'd never seen it before. The available mathematical data predicted that Neptune must exist, or else the equations governing the movements of the other, already known planets in the solar system, and Uranus in particular, would be incorrect.

a booklet, based on an earlier lecture, entitled "Radioastronomy and Communication through Space", in which he asked, "What can we talk about with our remote friends?" His assured answer was that "we have mathematics in common, and physics, and astronomy. We can open our discourse on common ground." Purcell envisaged an interstellar conversation "which is, in the deepest sense, utterly benign. No one can threaten anyone with objects. All you can do is exchange ideas."

Talking instead of listening

Frank Drake celebrated an upgrade of the **Arecibo radio telescope** on 16 November 1974 by firing a radio message towards the globular star cluster M13 some 25,000 light years away. M13 happened to be in the sky above the Arecibo dish. The message was broadcast for less than three minutes, and will take 25,000 years to arrive at M13. Then it'll be another 25,000 years before we can receive a reply – at least so long as our current assumptions about alien communications are correct.

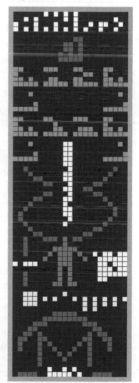

Drake's brief transmission was preceded by a prototype message, coded mathematically in a base-two counting system. This trial run was aimed at his human colleagues and transmitted by paper and old-fashioned postal mail. It took the form of precisely 551 characters, various sequences of ones and zeros ("on" and "off" pulses) in seemingly random order. Could any mind, let alone an alien one, make sense of it? Carl Sagan quickly worked it out – but he would, wouldn't he? The number 551 can only be divided by 19 and 29. Those two smaller numbers, in turn, are prime. They can only be divided by one, or by themselves. Prime numbers are fascinating to mathematicians. More importantly, they apply no matter what your counting system happens to be. Once you spot the prime number possibilities in the number 551, the natural temptation – natural to us, at

The picture formed by the correct arrangement of Frank Drake's binary Arecibo radio transmission.

any rate – is to make 29 rows of 19 pulses, or 19 rows of 29 pulses, and see what happens. After assigning a colour difference to the ones and zeros (black for the ones and white for the zeros, for instance) you get a picture.

In the bottom half of the picture, a humanoid figure can be discerned. The upper-right corner shows the simple relationship between carbon and oxygen atoms. Other clues in the prototype image equate the order of planets in our solar system with binary number combinations, and so on, in a similar fashion to the **Pioneer 10 plaque** (see p.152). The final version of Drake's Arecibo transmission took the form of 1679 bits, with the on–off pulses modulated by 10-hertz variations in the radio frequency. The resulting stream of bits can be arranged rectangularly as 73 rows in 23 columns, until it looks like the image on the previous page. In a 1974 press conference, Sagan described the resulting picture: "Here's the sun. The sun has planets. This is the third planet. We come from the third planet. Who are we? Here is a stick diagram of what we look like…" The pattern of bits was arranged in such a way that if one chose the alternative grid, 23 rows in 73 columns, no obvious picture emerged.

In the 1997 movie *Contact*, based on Carl Sagan's novel of the same name, an alien message is received, and turns out to be multi-dimensional rather than flat, and fiendishly harder to unravel than Drake's simple grid – but the aliens have still used mathematics to make their point. Galileo

Crowd-sourcing a message

In October 2009, a project called "A Message from Earth" (AMFE) directed a high-powered digital radio signal towards Gliese 581, the red dwarf star whose multiple planetary system has been on astronomers' minds so much in recent years (see p.168). The signal consisted of 501 separate messages selected through a competition on the social networking site Bebo, involving half a million of the site's users. The bundled transmission was sent using the State Space Agency of Ukraine's (SSAU) RT-70 radar telescope, and it will reach the vicinity of Gliese 581 in early 2029.

With a mix of digitally scanned photos, drawings and texts expressing the interests and hopes of Bebo users, AMFE lacks the mathematical rigour that one might expect from an interstellar messaging project. Dr Aleksandr Zaitsev, from the Russian Academy of Science's Institute of Radio Engineering and Electronics, is one of the principals behind AMFE. "I understand that in the majority of cases these messages may be naïve, but I also hope that we will receive a creative and fresh look at the subject", he told *The Guardian* newspaper. It's all just a bit of fun… isn't it?

amfe.lessrain.com

insisted that the universe is "written in the mathematical language, and the symbols are triangles, circles and other geometrical figures, without whose help it is impossible to comprehend a single word of it." Four centuries later, he's not been proved wrong, and we assume that alien intellects would also depend on mathematics, "without which", Galileo warns, "one wanders in vain through a dark labyrinth."

Interstellar television

Radio has been used on Earth for more than a century, but most signals have been too weak to escape our atmosphere. The various space agencies' communications with solar system space probes might be detectable by alien antennae, but they are too rare, and too feeble, to count for much. They're also extremely directional, aimed as precisely as possible at the probes to minimize wastage of already faint signals. It's unlikely that a listening alien's equipment would happen to be in radio line of sight. You might, therefore, be surprised that some of our most frivolous radio emanations, intended for strictly terrestrial audiences, are powerful enough to leak into space in all directions and traverse the galaxy.

The first episode of the popular US sitcom *I Love Lucy* was broadcast on 15 October 1951, and picked up by eager TV sets across the nation within a fraction of a second. After a full second, the signal would have left Earth, swept past the orbit of the moon and headed into deep space. And it's still going, at the speed of light. In our galactic neighbourhood, approximately 10,000 star systems have been exposed to *I Love Lucy*'s debut in the half-century since its first transmission. Primitive broadcasts dating back to Nazi Germany are a decade further ahead in the televisual race across the galaxy.

> "Television is ephemeral, a fact that some will find reassuring. But Earthlings will continue to pump the kilowatts into the ether. And eventually, when those signals have washed over a few hundred thousand star systems, someone may notice."
>
> Seth Shostak, senior astronomer, SETI Institute

Television is embedded in the fabric of our society, from *Star Trek* to washing powder adverts, *The Sopranos* to *The X Factor*. Abraham Loeb, professor of astronomy and director of the Institute for Theory and Computation at Harvard University, suggests that a radio telescope built to study distant galaxies might also be able to pick up the equivalent of alien TV transmissions, even if their content is meaningless to us. The **Low-Frequency Array** (LOFAR), an arrangement of simple tent-shaped wire antennae spread across Holland and Germany, can be tuned to frequencies under 250MHz, much lower than those favoured by most SETI radio searches, and closer to those in which most terrestrial TV is transmitted. If aliens are as addicted to electronic entertainment as we are, shouldn't we be able to spot their signals leaking into space, just as easily as they can tune into *I Love Lucy*?

One difficulty is that the radio emanations that carry radio and TV programmes are just that: carrier signals. The actual information content takes the form of relatively minor modulations in the signal, such as a rise and fall of the amplitude (the maximum height of each radio wave) or sudden starts and stops between pulses (as in your basic Morse code method). As the radio waves spread out through space, their energy dissipates and the information content becomes increasingly hard to discern, until all that can be detected is the faint presence of the carrier wave itself. Researchers at the SETI Institute claim that they would be hard-pressed to derive a meaningful text message from a mobile phone on the surface of Pluto, let alone a TV programme from a planet in another star system. According to senior astronomer Seth Shostak, "We would only be able to find TV signals comparable to ours from a distance of less than one light year."

Perhaps the most obvious problem for any radio-based approach to SETI is that the entire concept of radio technology looks as if it's about to be relegated to museums here on Earth. Why, then, do we imagine that aliens might be using it? In just over a century, we've moved from Morse signals and radio broadcasts, through the television era, into the realm of mobile phone and computer communications. The promiscuous crackle and buzz of our electromagnetic communications are being tucked away, ever more efficiently, into fibre optics and small, focused radio dishes that minimize the wasteful leakage of signals. The scattergun radio and TV towers of old are vanishing. This means that, over the next few decades, Earth will be heading towards radio silence at just the kinds of frequencies that any alien SETI radio experts might be searching. Similarly, we have to imagine that the noisy radio phase of an alien civilization could be just as transitory. Our attempts to listen for alien radio signals might fundamentally be a losing game.

As Frank Drake demonstrated in 1974, it takes a device the size of the Arecibo radio dish, pointed at a specific target in the sky, to transmit anything meaningful towards another star system. We have to wonder if an alien civilization would take the trouble, or expend the energy required, to text-message our solar system in such a way. Freeman Dyson, one of the great figures of mid-twentieth-century theoretical physics, thinks it unlikely. In his 1979 book *Disturbing the Universe*, he says that our galaxy "is sparsely populated and uncooperative. Intelligence is very rare, and none are interested in helping us discover them. Even under these unfavourable conditions, the search for intelligence is not hopeless" so long as we "turn aside from radio messages" and learn instead "how to recognize artificial objects".

Our gift to them: Pioneer 10

Launched on 2 March 1972, NASA's Pioneer 10 robot probe was the first spacecraft to obtain close-up images of Jupiter. Plutonium batteries kept the tiny craft operating far longer than anyone had anticipated. Its last detectable signal was returned in January 2003, more than thirty years after its launch. By then, the entire solar system was 13,000 million km in its wake. Mission planners always knew that Pioneer 10's trajectory would eventually fling it away from the sun and towards the stars – specifically, towards Aldebaran in the constellation of Taurus, 68 light years away. Pioneer won't reach Aldebaran for another two million years, but its ghostly shell might survive the trip: a fossil remnant of an earthly civilization perhaps long vanished by then. There is the slight chance that its artificial construction will be noted by an alien intelligence in the depths of space.

In the early 1970s, freelance British space journalist Frank Burgess chatted to Carl Sagan about the possibility of sending a visual message to aliens on the flanks of Pioneer, precisely because it was known from the outset that the little craft would one day escape our solar system. Intrigued, Sagan, Frank Drake and like-minded colleagues put the idea into practice. With permission from NASA, they added a small rectangular plaque to Pioneer, made from corrosion-proof gold and aluminium. It shows where our species lives, and our biological form. At the top left of the diagram is a schematic of two hydrogen atoms bound up in a molecule of H_2. The tiny energy shifts in hydrogen should be known to an alien civilization. These "hyperfine transitions" can then be used as a yardstick for both time and physical length throughout the universe. That gets rid of the problem of metric, imperial or other units of measurement that confuse us humans, never mind any aliens.

As a further size check, the binary equivalent of the decimal number eight is shown between tote marks indicating the height of the two human figures scaled against the spacecraft itself, which is also shown in line silhouette on the plaque. The radial pattern to the left represents the position of our sun relative to fourteen pulsars and to the centre of the galaxy. The binary digits on the other lines denote time. There is also a diagram showing Pioneer 10's journey, accelerating past the largest planet in our solar system with its antenna pointing back to its origin on the third planet.

Would *you* understand the plaque, as drawn by people of your own species? Perhaps not at first, but intelligent aliens might dedicate their brightest minds to the task of unravelling its mysteries, so perhaps they'll have better luck. For now, the plaque is better understood as a message to ourselves rather than to aliens. It signifies that we at least tried to head towards the stars, even while clashing ideologies and social problems afflicted us on Earth. It's a technological cave painting, created as much because we felt like it as for any other reason: a flattering portrait of ourselves as we would like to be seen.

Attitudes have changed since 1972, and we no longer share all of the assumptions made in the scientific symbols, the arbitrary ethnicity of the human figures and the woman's smaller size (and presumably,

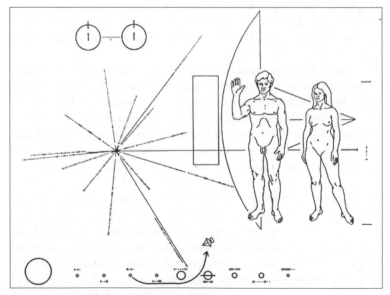

Greetings from Earth: the Pioneer 10 plaque.

Earth's greatest hits

The Voyager Golden Records were attached to the Voyager 1 and Voyager 2 spacecraft launched in 1977 to survey the solar system's outer planets. Each record was a 12-inch gold-plated copper disc engraved with a spiral of dips and dents, using the best phonograph technology available at that time, prior to the age of laser discs and DVDs. The protective case included a diagram of how the stylus (included in the kit) could be used to play the record, a Space-Age LP of Earth's greatest hits as chosen by Carl Sagan and his committee.

The record contained more than a hundred images of life on Earth, including diagrams of mathematical measurements, the solar system, DNA and conception. Photos featured the natural world (dolphins, an elephant, a toad, a snowflake and so on), and depicted humans from different cultures engaged in a variety of everyday activities: cooking, dancing, shopping, studying, playing musical instruments, sitting in traffic…

The audio selections included a variety of natural sounds, such as surf, wind and thunder, along with the songs of birds and whales, plus the artificial noises made by trains, tractors and aircraft. Music chosen to delight alien discoverers ranged from the classical works of Bach and Beethoven to Peruvian panpipes, a Pygmy girls' initiation song and Chuck Berry. Humanity also introduced itself in short greetings in 55 languages and written messages from President Jimmy Carter, UN Secretary-General Kurt Waldheim and other leaders from a civilization that, in all likelihood, will have vanished into dust by the time anyone out there finds those records.

status) compared with the man's. Why is the man waving, while the woman seems to be gazing submissively at him? Sagan's wife at the time, Linda, made the template drawings after taking advice from a number of scientists. Everyone concurred that if both the man and the woman held up their hands in greeting, the aliens would think that we all walked around all day with one hand in the air. It made sense to show a variety of postures, although it never occurred to anyone in 1970s NASA that the woman might do the waving while the man demonstrated the arms down position.

The plaque (and its duplicate, attached to the 1973 Pioneer 11 probe) was a physical embodiment of the radio message that Frank Drake beamed into space in 1974, when he celebrated the upgrade of the Arecibo radio telescope. Drake, and almost all SETI researchers since, have assumed that there's one language that intelligent aliens must share with us: mathematics.

The planet hunters

When Frank Drake wrote his famous equation in 1961 (see p.131), the existence of other planets orbiting other suns was just an assumption. Today it's an accepted fact. New astronomical techniques on Earth, allied to space-based instruments, prove beyond doubt the "plurality of worlds".

How do astronomers hunt for planets?

There are at least 200 billion stars in our galaxy, the Milky Way. The prospect that many of them must have planets seems fairly obvious. But those stars are staggeringly far away, so any planets they might have in their orbits are exceptionally hard to detect. Fortunately, astronomers don't have to look for planets directly. They can search instead for the influences that planets exert on what *can* be seen: the stars themselves.

Astronomers use a standard reference system, the **Hertzsprung-Russell diagram**, to categorize individual stars. The diagram plots the surface temperatures of certain classes of star against their brightness (luminosity), measured in terms of how much total energy they emit per second. The brightness of a star seems to dwindle the further it is away from us. By including corrections to account for that distance, the "absolute" brightness of a star can be established.

Most stars, including ours, fall into a narrow band within the Hertzsprung-Russell diagram, called the "main sequence". Stars falling outside this sequence range from old, dim red giants fading fitfully away at the end of their lives to dense, superhot blue giants, burning their fuels ferociously fast. Only one in a thousand stars are blue supergiants, but they are pretty obvious in the night sky. Their brightness compensates for their scarcity. The best known example is Rigel (in the constellation Orion), which is 20 times larger than the sun, and 60,000 times brighter. It's not impossible that blue supergiants, and other types of star very unlike our relatively peaceful sun, could harbour life-bearing planets. It's just that main sequence stars are more likely candidates, for the simple reason that one of them (and guess which one) has proved, beyond doubt, a tip-top contender.

The mass of a star in its prime of life determines what happens to it after its main sequence phase. Stars the size of our sun turn hydrogen into helium in their cores, but eventually the hydrogen runs out and the core can no longer generate enough outward radiation to balance the inward pull of gravity. The centre of the star shrinks while the outer layers expand. This makes the star cooler overall, and it becomes a red giant. Ultimately, the energy generation fades almost completely and the star collapses into a white dwarf. It's possible that white dwarfs may eventually dim yet further, to become black dwarfs, but that final fade-out would be so slow that we may never be able to prove that it happens.

That's the average stellar biography. But if a star is one of those comparatively rare examples more than ten times the size of our sun, a different fate lies in store. It will finish its lifetime with a sudden collapse of its core. The protons and electrons in its atoms undergo a terrible transformation, leaving a core composed entirely of neutrons and exerting an immense gravitational field. The outer parts of the star are hurled away in a violent explosion, and the star becomes a supernova. Large stars tend to end their lives as supernovae.

At the opposite end of the scale, stars that started life very much smaller than our sun may just dim over time to become unremarkable brown dwarfs, emitting so little light as to be almost invisible to our telescopes. Planet hunters focus largely – though by no means exclusively – on middle-aged, modest-sized, well-behaved stars similar to our sun, which is classified as a **yellow dwarf star**. These are not rare, nor hard to find. Approximately one star in ten is very like our sun.

Discs of space dust surround many main sequence stars, and this dust can be detected because it absorbs some of the starlight and re-emits it as slightly less energetic infrared radiation. Smudgy, dense-looking features in the discs often suggest the emergence of protoplanets. The Hubble Space Telescope (HST) has captured many images of **protoplanetary discs**: the embryos of future solar systems. The discs are, by definition, similar in diameter to an entire solar system, and therefore relatively easy to spot. Finding actual planets is tougher, because they are usually too

Embryo of a future star system: a protoplanetary disc of gas and dust swirls around a brown dwarf star (artist's concept).

small to observe, even in Hubble's powerful field of view. Confirmation of planets in other solar systems (extrasolar planets, or exoplanets) calls for subtle measurements of candidate stars.

Astrometry: measuring the position and motion of stars

As a planet orbits a star, the star also moves in its own small orbit around the system's collective centre of mass. Remember Otto Struve's work on stellar rotation, and that skater with her arms outstretched (see p.119)? Now imagine a large, heavy man in white coat and tails, spinning on his skates. As the spotlight catches him, he seems to be the only performer on the rink. Gradually, you notice he's struggling to spin on one particular spot. He's wobbling, and his skates are gouging a small circle in the ice. You wonder what might be causing this imbalance. If you assume that he's holding onto a very much smaller and lighter companion in a dark costume, and the pair are dancing an arm's length apart, you'll realize that he's being pulled slightly away from his centre of gravity by her small, yet still significant, mass.

Now imagine looking at this scene while lying on the ice, perhaps trying to take low-angle video footage of the action. The big dancer would seem to move from side to side, and you'd notice him veering towards you or away from you during the most easterly or westerly phases of each side-to-side sweep. That's because his circular progress does indeed bring him alternately closer to, and further from, you. In the astrometry technique, a star's position in the sky is logged very precisely. Any observed side-to-side deviations can be explained by the gravitational influence of orbiting planets.

Unfortunately, the deviations caused by the puny pull of planets are extremely difficult to detect. This is because the off-centre motions of the star are so small, they amount to just a fraction of the star's diameter. (Astrometry has, however, been used for many decades to help identify the larger movements in binary star systems, where even the smaller partner tends to be substantially more massive than any planet.) The next, and more precise, planet-hunting method uses the astronomical equivalent of a police speed trap radar.

Gravitational lensing

Most branches of science treat gravity as an invisible force exerted by all objects in the universe. Relativity treats gravity not as a force but as a distortion in space-time caused by the presence of the most obviously massive objects: galaxies, stars and planets. When light travels through space-time, it always does so in a straight line. It's not the light path, so much as space-time itself, that is distorted by mass.

As a star drifts through space, stars immediately adjacent to its disc appear, from our point of view, to shift in the sky. This is because every star distorts the space-time around it. The effect is to create a "gravitational lens", a bobble-like zone of space around a star that bends light from background stars. Just like any other convex lens, this can have the effect of magnifying the background stars. The "images" that we can detect this way are usually very smeared, but subtle investigation can hint at the presence of exoplanets orbiting the background stars.

Chance alignments of nearby stars, one precisely behind the other, are exploited for this kind of research. As the Earth moves through space, and our viewpoint changes, occasional alignments occur, and last for a few days, before the Earth's continuing movement breaks the spell. Only a few exoplanets have been found in this way, which is especially sensitive to planets orbiting at large distances from their stars.

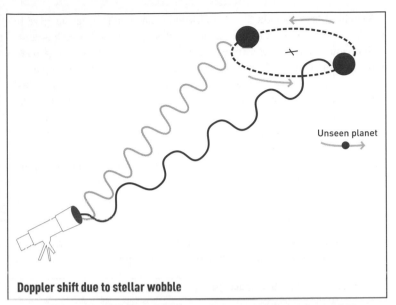

Doppler shift due to stellar wobble

A planet's gravity pulls its star off-centre. Doppler shifts betray the wobble when the starlight's wavelengths shorten as the star approaches us and lengthen when it recedes.

The radial velocity method

Before we can get to grips with how astronomers use the "radial velocity" method to look for exoplanets, we have to understand a phenomenon known as **Doppler shift**, named in honour of the Austrian physicist, Christian Doppler (1803–53), who described it in the 1840s. This is the change we observe in the frequency of a wave if we're moving relative to the source of the wave, or if the wave source is moving relative to us. Redshifts, as used by Edwin Hubble to prove that galaxies are rushing away from each other (see box on p.125), are just Doppler shifts going in a particular direction – towards the red end of the spectrum.

Doppler's discovery applies just as well to sound waves. Think of a lovely old-fashioned locomotive hurtling towards you, tooting its shrill whistle like a banshee. As the train rushes past, the whistle's sound changes its character, from high-pitched screech to a slightly lower tone. What's happening is that the train's rapid movement away from you stretches the sound waves, so that by the time they reach your ears, they are at a lower frequency.

Switching back to light (EM) wavelengths, and redirecting our gaze to the sky, we learn that exoplanet hunters analyse the redshifts and blueshifts of individual stars to determine whether or not those stars have planets. From now on, we'll use Doppler terminology when talking about exoplanets and the radial velocity method. Extremely small variations in a star's radial velocity – the speed with which it moves towards or away from Earth – can be detected by monitoring Doppler shifts in its light spectrum. It was Otto Struve, the pioneering expert in stellar radial velocity, who proposed (in 1952) that sensitive spectrographs could be used to find exoplanets in this way. Now we have to spend a few moments on this business of spectrographs. They are as basic to an astronomer as spanners to a car mechanic.

Analysing starlight

When sunlight is directed through a glass prism, the colours spread out into a familiar rainbow, or spectrum (the plural is "spectra"). A

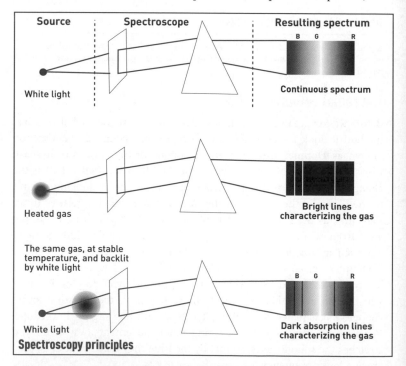

Spectroscopy principles

Bright vertical bars are the light signatures of particular chemicals when they're hot. When those same bars are dark, the chemical is cool, and blocks light from behind.

spectrograph is a box of mirrors, prisms, lenses, diffraction gratings and sensors, attached to the end of a telescope, that captures and records spectra in extreme detail. A star's spectrum has a barcode like pattern of fine, vertical lines. The dark absorption lines, known as **Fraunhofer lines**, represent absences of light. They are caused by certain chemicals in a star's outer atmosphere, or in surrounding interstellar dust clouds, absorbing some of the frequencies of light hurled out by the visible outer "surface" of the star's glowing ball (the photosphere). The vivid bright emission lines show particular chemical elements glowing fiercely in that star. The barcode pattern of absorption and emission lines shifts towards the blue or red end of the spectrum whenever Doppler shifts come into play.

Every scientific instrument needs to be calibrated. What use is a measurement unless you have something to compare it against? A ruler, marked off in inches or centimetres, has been calibrated against internationally agreed definitions of what a centimetre means, or how long an inch is. Similarly, the absorption or emission lines in any kind of spectrum don't mean much without calibration. Over the last century or so, chemicals of countless kinds – from pure elements to complex molecules – have been heated up in laboratories until they glow. (The substances are heated, not burned, because burning is a chemical reaction, which would muddy any results.) Each substance's light spectrum has been measured and catalogued, and astronomers can compare the results of their stellar observations against this vast list of known spectra. Tricky work, even in the controlled circumstances of a laboratory. It's remarkable, then, that ninety years ago, a young woman discovered what the sun is made of – it's not quite so remarkable that, at first, her wise male scientist colleagues didn't believe her…

Cecilia Payne and the composition of stars

In 1905, a five-year-old English girl, **Cecilia Payne** (1900–79), saw a meteor shooting across the sky and decided to become an astronomer when she grew up. Fourteen years later, as a student at Newnham College, Cambridge, she attended a lecture by famed astronomer Arthur Eddington, in which he explained how his solar eclipse observations had proved Einstein's thesis that massive objects such as the sun bend the light from distant stars. Payne tried to get access to the university's telescope, only to find its guardian rushing off to warn his superiors, "There's a *woman* out there, asking questions!" Payne's career was dogged by sexism, yet she made one of the most important discoveries in the history of astronomy.

At the age of 23, Payne moved to Harvard University, Massachusetts, where she analysed the sun's spectrum and found that the entire astronomical community was misreading it. The prevailing theory was that stars must be superhot balls of iron with luminescent outer shells of gaseous iron. Payne told a different story. In a brave academic paper, "Stellar Atmospheres", published in 1925, she asserted that the sun was almost all hydrogen, with only small fractions of iron and other chemicals in its make-up. Her male critics were outraged, because they knew, as no foolish woman ever could, that if the sun were made of hydrogen, it would burn out in a cosmic flash rather than shining on, as it so obviously does, for countless millennia.

Eventually, Payne's work was replicated by other scientists and she was heralded as brilliant. Now the path was clear to discover how the sun, and all other stars, generate vast amounts of energy over billions of years. It's a nuclear process, not one of fire and burning. After Payne's breakthrough, the sun was no longer quite so mysterious. A catalogue of the chemicals in its photosphere and atmosphere was assembled using the same technique that she used: the careful and laborious comparison of bits of the sun's spectrum with glowing chemicals in the laboratory. In just the same way, we now analyse the composition of more distant stars. But the chemical nature of stars is an incidental part of what exoplanet hunters are looking for. It's not molecules but motions that they seek. They look for extremely fine Doppler effects, the slight **left–right shifts** of those chemical barcodes in the spectra of stars, caused by a star's slight wobble.

How spectrographs find exoplanets

The spectra of distant galaxies racing away from us at colossal speeds are noticeably redshifted, but the light from any star in our neighbourhood of the galaxy is barely shifted at all, because – from Earth's perspective – such stars are practically motionless from day to day, year to year. In the 1990s, US astronomers Geoff Marcy and Paul Butler found a method of identifying spectral shifts caused by a star wobbling off-centre at the rate of just a few metres per second. The shifts from such an insignificant motion would be too small for any conventional spectrograph to pick up, so they spent six years building a new one. They did so in conjunction with Steve Vogt, who had already built a spectrograph for the Lick Observatory near San José, California.

Vogt's instrument was not simply a box plugged onto the seeing end of a telescope. It was a super-clean and utterly black room-sized chamber, into

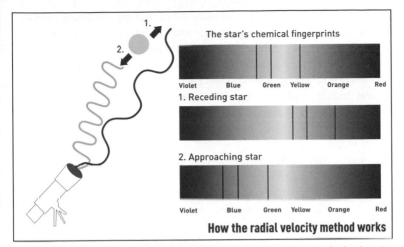

How the radial velocity method works

The characteristic chemical "bar codes" of a star's spectrum show regular back-and-forth deviations, matching the gravitational influences of planets orbiting the star.

which light from one of the Lick telescopes was diverted. The resulting spectra were extremely sharp and detailed. They had to be, in order for sensors to pick out superfine Doppler shifts and other data. Vogt could spot Doppler shifts down to a few tens of metres per second. Marcy and Butler needed to get that down to around twelve metres per second, for that's the tiny effect that even a planet the size of Jupiter exerts on its parent star.

Two immense technological challenges had to be solved. The first was to simultaneously compare the spectrum of a target star with a terrestrially sourced chemical spectrum, because, when it comes to monitoring a Doppler shift that might take just half an hour to cycle back and forth, the relevant telescope has to remain pointed at the star throughout the observation. Not a single second of observation time can be wasted cross-checking with the reference spectra. Both processes must overlap, so that real-time comparisons can be made between both spectra.

Just as Payne and her successors have done for a generation, Butler and Marcy searched through laboratory catalogues of glowing chemicals, looking for one whose barcode lines were scattered fairly widely across the whole light spectrum, from the blue end to the red. They found one that fitted the bill – iodine – which creates many lines, literally across the spectrum. They then built an iodine cell, a transparent chamber full of the gas, and placed it between the telescope and the spectrograph. To cut a long and incredibly complicated story short, they created a neat overlap between a target star's iodine signature and their earthly iodine's equivalent signature. If the overlaps slid back and forth out of register at regular intervals, they could be reasonably sure that a star was Doppler shifting.

The second challenge was that the Earth itself, ever restless in its orbit around the sun – and turning on its own axis to boot – adds a tremendous amount of unwanted clutter to a Doppler observation. In 1991, British astronomer Andrew Lyne, working at the huge Jodrell Bank radio telescope in Cheshire, northwest England, thought he'd found a planet orbiting a pulsar. He published his results in the science journal *Nature*, and looked set to be one of the earliest discoverers of an extrasolar planet. A short while later, he addressed a conference of his scientific peers and told them, "I must issue a retraction and make an apology". His numbers were wrong. He hadn't found a planet after all, and just wanted to sink through the floor with shame. In fact, he got a thunderous round of applause as his audience saluted his scientific honesty and personal courage.

Lyne had used a radio-based planet detection method known as **pulsar timing**. Pulsars, the dense and fast-spinning remnants of supernova explosions, don't make ideal stars for life-bearing planets. That's a pity because a pulsar's extremely regular lighthouse-style beams of radio energy are easy to monitor for the tiniest Doppler variations. If a pulsar is wobbling, it's because a companion object is tugging at it. The first confirmed discovery of an exoplanet was made using this method, which wasn't originally designed for planet-hunting, but is so sensitive it can detect planets less than a tenth the mass of Earth. Unfortunately, that first confirmation wasn't Lyne's. He had accounted for the Earth's motions through space, but had approximated some of his numbers for convenience, basing his calculations on a circular orbit for the Earth. Indeed, it's orbit is very nearly circular – but not quite. In his excitement, Lyne forgot to make that one, last tiny correction to his equations.

Butler was keen to avoid this kind of disaster. He wrote a computer program that took into account the distorting effects of the Earth's atmosphere, the motions of the Earth through space, its day–night rotation on its own axis, a number of unavoidable ambiguities introduced (despite Vogt's best efforts) by the Lick spectrograph's lenses and mirrors, plus dozens of other factors. It seemed such an unwieldy program, many astronomers laughed at it. But by 1995, after six years of development, it was finally working. Butler has described it as, "My Rembrandt, the closest to great art that I'll ever get".

Telling a swell from a wobble

Doppler shifts are generated by an exoplanet's pull on its parent star, but there is another source of shifts that can come from the star alone. Some stars expand and contract rhythmically as their internal nuclear fusion pressures increase and subside at regular intervals. The outer layers of the photosphere correspondingly move towards the Earth and away from it, almost exactly emulating the small Doppler shifts that would be generated by an exoplanet orbiting nearby. If a star swells up slightly and then contracts again, its surface area changes, and its overall luminosity flickers up and down: a sure sign that this star can't be trusted by exoplanet hunters. On the other hand, if a star's light intensity remains absolutely stable while it exhibits Doppler shifting, then exoplanets or other gravitationally significant companion bodies are most likely to be responsible for the shifts.

The first exoplanet discoveries

The first announcement of a planet orbiting a star other than our own sun (an exoplanet) was published in 1988 by Canadian astronomers Bruce Campbell, G.A.H. Walker and S. Yang. They made the discovery using the radial velocity method. The star in question was **Gamma Cephei**, one of the partners in a binary star system approximately 45 light years away in the constellation Cepheus. There was some argument that the supposed exoplanets might instead be brown dwarfs, gaseous objects larger than planets but smaller than stars. It took some years for the wider astronomical community to accept their findings.

Meanwhile, in 1992, radio astronomers Aleksander Wolszczan and Dale Frail, using the Arecibo Observatory in Puerto Rico, found two planets orbiting an energetic, fast-spinning pulsar-type star, designated PSR 1257+12. Their discovery was swiftly confirmed, and is generally considered to be the first definitive detection of exoplanets. A pulsar's planets may be formed from remnants of the supernova that produced the pulsar in the first place, or they may be the rocky cores of planets that survived the supernova. Either way, they are unlikely abodes for life.

> **"After the discovery of 51 Pegasi, everyone wondered if it was a one-in-a-million observation. The answer is 'no'. Planets aren't rare after all."**
>
> Paul Butler, astronomer, 1996

Of more interest to us is **51 Pegasi b**, an extrasolar planet approximately fifty light years away from our sun, in the constellation Pegasus, and the first planet to be discovered orbiting a sun-like star. The "b" indicates that the planet was the first one discovered orbiting that star. Further planets associated with any star would be designated c, d, e, f and so on. 51 Pegasi b's discovery was announced in October 1995 by Michel Mayor and Didier Queloz in *Nature*, on the basis of radial velocity measurements made by the Observatoire de Haute-Provence in France. They used a technique not unlike Butler, Vogt and Marcy's, only with somewhat simpler equipment.

After spending so many years gearing up the Lick spectrograph, the iodine cell trio was dismayed by Mayor's announcement. They'd been beaten to a potentially colossal discovery. They also felt that Mayor's spectroscopic equipment couldn't be as accurate as theirs, so they set out to prove the French duo wrong. What they found was that Mayor and Queloz's planet was so huge, and was orbiting so close to its star, that the Doppler shifts were actually quite easy to spot. 51 Pegasi may be sun-like, but planet "b" is certainly not Earth's twin. It's a super-massive Jupiter-class world, orbiting extremely close to its parent star.

In 1996, George Gatewood at the University of Pittsburgh announced the existence of a Jupiter-sized object circling the fourth nearest star to us, Lalande 21185, six light years away in the constellation Ursa Major. The giant planet's orbital period (or year) is approximately six Earth years, and the gravity equations pointed to the existence of other, smaller planets in the system as a whole. "The periodicity and the mass distribution overall looks much like the sort of thing we'd find in our own backyard", Gatewood reported.

But some backyards are stranger than others. A dazzling array of weird worlds have been found orbiting stars quite unlike our sun. The planet orbiting one of the three stars in the 16 Cygni triple-star system has an eccentric, egg-shaped orbit like that of a careening asteroid, yet its mass is greater than Saturn's. More intriguing is **GJ 1214 b**, a planet six times heavier than Earth, with three times its diameter. Some researchers, including its discoverer, Harvard astronomer David Charbonneau, believe that it's made mostly of liquid water. It orbits a red dwarf, a relatively small and cool star about a third of the diameter of ours, located forty light years from Earth in the constellation Ophiuchus. The planet is a hundred times closer to its star than the Earth is to the sun, so GJ 1214 b is not, at first glance, a Goldilocks world. Although its star obviously hasn't scorched away the planet's water, its deep oceans must be very hot. A thick, dense and possibly hydrogen-rich atmosphere may shield the water from the star's fearsome blaze.

Yes, it is possible to work out some qualities of an alien sky, even if arguments remain about its exact chemical composition. In 2008, astronomers using the Hubble Space Telescope's Near Infrared Camera and Multi-Object Spectrometer (NICMOS) detected carbon dioxide in a planet's atmosphere by subtracting the parent star's spectroscopic data from the combined light of star *and* planet. What was left was the planet's spectrum alone, composed of the starlight that it reflects back into space, minus those parts of the spectrum that its atmosphere absorbs. Carbon dioxide is an interesting discovery, even though the planet in question, HD 189733 b, is a hot, Jupiter-sized gas giant orbiting too close to its star to be habitable. Carbon dioxide is probably

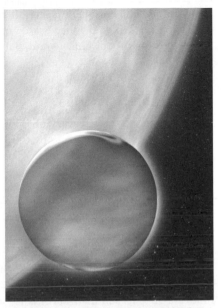

Exoplanet HD 189733 b, discovered by the Hubble Space Telescope in 2008, orbits too close to its parent star to be habitable (artist's concept).

quite common in planetary atmospheres, though. Mars is swathed in a thin veil of the stuff, while Venus swelters under a dense layer of it. What we'd really like to discover is an exoplanet with an oxygen-rich sky.

Perhaps we should try looking for blue exoplanets?

In 2010, Lucy McFadden, a researcher at NASA's Goddard Space Flight Center, investigated colour information from planets in our own solar system as seen by cameras on space probes hurtling towards distant targets (such as the Deep Impact probe en route to its close encounter with Comet Hartley 2). Earth, in particular, stood out, appearing much bluer than Mars, Venus, Mercury or the gas giants. According to McFadden and her colleagues, exoplanets that look blue may be the interesting ones.

This assumption sounds simplistic, but there is good science behind it. Earth's unique colour, when seen from a distance, is the product of two factors: the scattering of blue light by dust and water vapour in its atmosphere, and the fact that it doesn't absorb much infrared light. Our air is low in infrared-absorbing gases such as methane and ammonia, in stark contrast to the gas-giant planets. If an exoplanet has a similar colour profile to Earth's, it doesn't necessarily mean that it has the cloudy skies and deep oceans of our home world, but it does mean we'll want to take a closer look.

Finding Earth-like worlds

When it comes to the discovery of a potentially *habitable* exoplanet, the first compelling case was made by the Lick Observatory, California, in September 2010. A team headed by Steve Vogt and Paul Butler found a planet, designated **Gliese 581 g**, just twenty light years away in the constellation Libra, orbiting the red dwarf star Gliese 581 (a frequent target for astronomers' attention in recent years). Earth-based telescopes detected the signatures of six planets orbiting the relatively cool sun, with two apparently close to the Goldilocks Zone.

This version of the zone isn't quite what we'd recognize from our terrestrial example, though, because the parent star is cooler than our sun. A year on Gliese 581 g lasts 37 Earth days. The planet is ten times closer to its sun than Earth is to ours, three or four times Earth's size and probably has a stronger gravity field. This still places it within the "Earth-like" category. On the downside, Gliese 581 g is so close to its parent star, tidal forces are likely to have halted its rotation around its axis, meaning one side always faces the sun and the other is perpetually dark (like our moon). Perhaps life thrives in the eternally locked grey margins between night and day. If this data holds up to further scrutiny, Gliese 581 g widens the potential envelope of the Goldilocks Zone.

"Earth-like properties are a little bit mysterious, but we have some ideas. You want water in liquid form, you want stable temperatures over the course of millions, preferably billions, of years so that Darwinian evolution can get a good toehold. You probably want a moon to stabilize the spin axis."

Geoff Marcy, astronomer, 2011

By November 2010, five hundred extrasolar planets had been discovered, though the nature of the business – from initial suspicions to confirmation – meant that no one could claim to have found *the* five hundredth planet on any given day. However, the *Extrasolar Planets Encyclopedia*, a database compiled by astrobiologist Jean Schneider of the Paris-Meudon Observatory, announced that the milestone was reached on 19 November, with the count rising to 502 a couple of days later.

Radial velocity measurements via Doppler analysis work well up to the level of quite large planets. Worlds that are actually as small as ours have only a negligible effect on their parent stars, and these can be hard to detect. Earth-sized planets have to be located with a different technique, known as the **transit method**. Whenever a planet crosses in front of (transits) its parent star's disc, it blots out a tiny fraction of the star's light. If the starlight dims and brightens again at specific intervals, it's reasonable to assume that an orbiting planet is responsible. Subtle variations in the transit times of one planet can be used to detect others in the same system. Short-term observations may just yield a mishmash of conflicting data, but over the long term, it is possible to work out the relationships between the target star and multiple planets. In 2011, the next great planet-hunting exercise – conducted by a NASA spacecraft using the transit method – left that 502 figure far behind.

The Kepler mission

NASA's Kepler spacecraft was launched in March 2009 from Cape Canaveral Air Force Station, Florida, to orbit the sun, trailing slightly behind the Earth. Its on-board telescopic instruments were designed to

Before beginning its planet-detecting quest, Kepler is placed on a stand for fuelling inside the hazardous processing facility at Cape Canaveral.

detect planets as they transit their stars, blocking a tiny fraction of the starlight from view. From May to September 2009, Kepler kept watch, unblinkingly, over 156,000 stars between the constellations Lyra and Cygnus. This target field was selected because it's rich in sun-like stars, and also far enough north of Earth's orbital plane (the ecliptic) that the dazzle of our own sun won't fog Kepler's view.

At the focus of Kepler's optical array, 42 charge-coupled device (CCD) panels comprise the business end of an extremely sensitive digital camera. A typical upmarket digital SLR camera has around 18 million pixels in its sensor; Kepler has more than 95 million. Their task is to look for extremely slight variations in light levels from each star, monitoring all 156,000 simultaneously, and refreshing the measurement data at half-hourly intervals. The effects of a transit last from about an hour to half a day, depending on the planet's orbit and the type of star. An odd dip in the light levels now and again isn't sufficient to confirm the presence of a planet; the dips have to repeat on the kind of neat cycle that we'd expect from a planet in a stable orbit passing in front of its parent star at regular intervals. A series of transits must exhibit similar depth (dips in starlight), duration (transit time) and period (time elapsed between successive transits) before a planet's presence can be confirmed. The more often a transit can be captured, the better the data gets.

The results of Kepler's quest were published in February 2011. It found 1235 potential transits. That doesn't mean 1235 individual stars have planets in their orbits. Some hog an unfair share, with at least 170 stars hosting entire solar systems. Of the planets tallied so far, around half must be at least the size of Neptune, or around 50 times the size of Earth, but 288 of them are no larger than 10 times the size of Earth. Heading towards the juicy end of the Drake numbers, 54 planets occupy the Goldilocks Zone, the potentially habitable orbital realm that's near enough to the parent star for liquid water to exist without turning to steam, but not so far away that it freezes. Of those 54 Goldilocks planets, 5 are the size of Earth.

This may sound like a disappointingly low number from a sprawl of 156,000 stars, but Kepler's static field of view covered only about a four-hundredth of the sky. Extrapolating from this – and bearing in mind the Mediocrity Principle (see p.29) that says this 1/400th fraction of the galaxy shouldn't be that different from any other four-hundredth – a simple calculation says that if a telescope of the future were to make an all-sky survey, it should find approximately 5 Earth-sized planets in life-friendly orbits in around 400 different patches of the sky. That's 2000 Earth-sized Goldilocks planets in our galaxy alone.

Next comes another scaling up of the numbers, because there's something we haven't yet mentioned: transits can only be observed when a star's planetary system happens to be correctly aligned with the telescope's line of sight. To use an analogy that might have pleased Aristotle, think of each solar system as a transparent crystalline disc, with the star in the middle and the tiny ant-like planets glued around it. From Kepler's viewpoint, the planets only pass directly in front of their suns if the disc is seen exactly edge-on. Only a small fraction of extrasolar systems can be expected to have such a convenient alignment. Big fat Jupiter-like gas giants orbiting very close to their parent stars may block some of the light even when the alignments are slightly askew. But for a planet in an Earth-sized orbit, the chances of it blocking any light are fewer than one in a hundred. That means we get to bump up our tally of Earth-like planets in the galaxy a hundred times. Now the magic figure is more like 200,000.

Time for one last jump in the maths. Kepler can only monitor stars that are local to our patch of the galaxy. Any stars further than around 3000 light years away are so small in its field of view, it can't reliably measure light variations. Remembering that the Milky Way is some 100,000 light years in diameter, it's obvious that we can only monitor a tiny fraction of what's out there. Some of the slots in the Drake equation are starting to fill up. According to the SETI Institute's Jill Tarter, a realistic extrapolation of the Kepler numbers "gives us the statistic that we can expect 50 billion planets in the galaxy, and 500 million of those are likely to be habitable". Whether or not any of them actually are inhabited is another matter.

Kepler's numbers at a glance

Number of stars monitored	156,000
Number of planet candidates identified	1235
Solar systems with multiple planets	170
Planets larger than Jupiter	19
Planets the size of Jupiter	165
Neptune-sized planets	662
Planets up to twice Earth's size	288
Planets similar in size to Earth	68
Planets in the Goldilocks Zone	54
Earth-sized planets in the Goldilocks Zone	5

The next-generation space telescope

The Kepler space telescope looks for the all-important transit clues that signify the existence of an extrasolar planet, but it's not a large enough instrument to magnify any exoplanets and give us a closer look at them. The Hubble Space Telescope (HST) – launched in 1990 and maintained by several astronaut visits since (most recently in 2009) – still packs a tremendous visual punch. Eventually it will start to fail. And there is no longer any prospect of repairing it, because NASA's fleet of space shuttles has been retired.

NASA is building an even more powerful replacement. The **James Webb Space Telescope (JWST)**, named in honour of a former NASA chief administrator, is expected to launch in 2014 on a European heavy-lift rocket. JWST is an infrared telescope with a 6.5m primary mirror mounted on top of a sunshade the size of a tennis court. The supporting spacecraft, with propulsion, communication equipment and solar power panels, juts out from the bottom of the sunshade. Dr Pam Sullivan, lead manager for JWST's Integrated Science Instrument Module, or ISIM, explains the need for such a large primary mirror: "It will be looking for very distant and faint galaxies, and often will pick up no more than a single photon of light per second from a target, so we want to catch as many photons as we can."

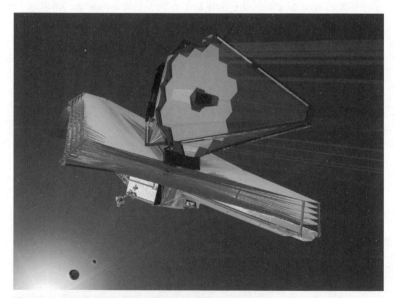

The James Webb Space Telescope, with its large primary mirror above a multi-layered sunshade designed to block infrared radiation (artist's concept).

This extreme need for sensitivity requires JWST to fly a special orbit, 1.5 million km away from Earth, known as L2, where the combined gravitational influences of the sun and Earth ensure that JWST is kept well away from the infrared "noise" of the Earth–moon system, and the glare of the sun itself. The spacecraft's huge sunshade blots it out. Hubble-style astronaut repair missions won't be possible that far out in space, so JWST will just have to fix its own problems. The telescope's primary mirror is divided into eighteen separate hexagonal panels, each about the size of the door on a pick-up truck. Each panel can be finely adjusted by remote control to adjust the focus.

JWST will look deeper into the universe than ever before, using infrared frequencies. Space is suffused with dust clouds, the ancient wreckage of stellar explosions and other cosmic events. These are the essential building materials for new stars, but they also block out much of the light from any objects behind them. Fortunately, the longer wavelengths at the red end of the spectrum can get through the dust.

JWST will also look further back into the past than ever before. No, it isn't a time machine. The further away a galaxy is, the younger it is, because of the time it takes for its light to reach us. At extreme distances, the light from those galaxies is redshifted to such an extent, there's almost nothing left at the visible end of the spectrum. Hubble can see galaxies in visible light, dating from as long as ten billion years ago, but JWST will collect extremely redshifted light dating from soon after the birth of the universe, something that's never yet been accomplished.

Closer to home, within our own galaxy, JWST will attempt to capture images of planets orbiting nearby stars, and try to determine whether any of them might be capable of sustaining life.

Intelligent civilizations

What happens if SETI achieves its aim and detects a signal from an alien intelligence, or finds some other incidental sign that "they" exist elsewhere in our galaxy? What if alien envoys actually land on Earth? Would we share any common cultural instincts? As we speculate about these momentous possibilities, we come up against some disturbing questions about ourselves and the possible fate of our own so-called civilization.

What should we do if E.T. calls?

In 1996, **John Billingham** (one of the instigators of NASA's shelved Project Cyclops proposal) and Michael Michaud (a fellow member of the International Academy of Astronautics) compiled a draft document outlining the need for a SETI post-detection protocol. This effort was supported by the Paris-based International Institute of Space Law.

Everything that happens in space at the behest of humans is subject to law. It's not a complete free-for-all up there. In the tense dawn of the Space Age, satellites and humans alike were hurled into the sky on rockets primarily designed for delivering megaton nuclear payloads across continents. The United Nations Committee on the Peaceful Uses of Outer Space (COPUOS) was established in 1958 in a bid to prevent an extension of warfare into the orbital realm or beyond.

Since then, a series of international treaties has limited the deployment of nuclear weapons in space, and – less successfully – sought to protect celestial bodies, such as the moon, Mars and asteroids, from being claimed by the first nation or corporation to make a grab for them. Ownership of potential space wealth is a theme that pays the lunch bills for many lawyers around the world, while yet more armies of lawyers deal with obscure aspects of international space communications law, such as the fair distribution of valued orbital slots, and so on. Space law counts for something.

With all this in mind, Billingham and his colleagues considered the legal implications of receiving a signal from an alien civilization. They wondered who might be in a position to spread the news, or to authorize transmitting a reply? When Frank Drake celebrated the 1974 upgrade of the huge radio telescope at the Arecibo Observatory, Puerto Rico, by shooting a radio message at a distant star cluster (see p.145), some critics were disturbed that he had been allowed to transmit anything at all.

Post-detection protocols

Billingham thinks that announcing our presence via radio should not be done lightly. In his 1996 position paper he wrote: "Clearly, sending a message to another civilization is more than just a scientific research project; it is a policy question that should be addressed by policy bodies." Given the lack of any other sensible option, he proposed the involvement of the United Nations, "the most universal of existing international policy bodies". Billingham knows the UN have more pressing worries than SETI, but who can doubt they'd be keen to get involved in a post-detection scenario if the situation arose?

Signal received: how should we respond if we detect an alien signal, like Jodie Foster's character, Dr Eleanor Arroway, in the movie *Contact* (1997)?

As we might expect from a sprawling international bureaucracy, the official positions on SETI are vague. A formal post-detection protocol has yet to be agreed by the UN or any government agency, but the International Academy of Astronautics (IAA), in conjunction with the International Institute of Space Law, came up with its own set of principles in 1989. It also established the **SETI Post-Detection Taskgroup** in 2005. This eclectic part-time team comprises about two dozen people, including two lawyers, representatives of the international media, SETI Institute scientists Jill Tarter and Seth Shostak, William Stoeger of the Vatican Observatory, Britain's astronomer royal Martin Rees and science-fiction writer Stephen Baxter (regarded by many as a successor to Arthur C. Clarke) The group's occasional meetings are chaired by quantum physi cist and cosmologist Paul Davies, a professor at Arizona State University.

The post-detection guiding principles are:

▶ Any individual or research institution (the discoverer) thinking it has detected a signal from an extraterrestrial intelligence should seek to verify that this is the most plausible explanation. If the evidence is fuzzy, the discoverer shouldn't claim anything beyond the fact that an unusual astronomical radio source has been located.

▶ The discoverer should inform his or her relevant national authorities, and other independent researchers and observatories, plus a long list of other relevant authorities, via the Central Bureau for Astronomical Telegrams of the International Astronomical Union (the standard formal route for all astronomical discoveries).

▶ The secretary general of the United Nations should be informed in accordance with Article XI of the "Treaty on Principles Governing the Activities of States in the Exploration and Use of Outer Space, Including the Moon and Other Bodies".

▶ A confirmed detection of extraterrestrial intelligence should be disseminated openly through scientific channels and public media. The discoverer should have the privilege of making the first public announcement.

The UN is aware of this protocol, and while it hasn't been enshrined into international law, most serious SETI researchers accept it as a useable framework. What can't be guaranteed is that everyone will stick to it if the big day comes. One way or another – and probably via Twitter – we'd all learn about the discovery of an alien signal pretty quickly. No bona fide

SETI researcher anywhere is interested in government conspiracies or deliberate attempts to hide alien signals or artefacts.

One hint of paternalistic secrecy is worth mentioning, however. As chair of the Post-Detection Taskgroup, **Paul Davies** supports an open policy, but in *The Eerie Silence* (2010), he questions the wisdom of letting everyone know, instantly, about any alien message that contains specific information content:

> "The implications of simply receiving such a message would be so startling and so disruptive that, although eventual disclosure is essential, every effort should be made to delay a public announcement until a thorough evaluation of the content had been conducted."

It's the only dubious sentiment in an otherwise liberal-minded and joyous book. Davies falls into the trap of thinking that some human beings are better prepared to cope with a given piece of news than others, although he does say it's best to try to keep governments out of the SETI process. But the global village has ears everywhere. He "cannot imagine how the scientists involved would be left in peace" while they crunch the data, and concedes that "keeping the lid on such a discovery would present enormous obstacles."

Assessing the impact of an alien signal

The Post-Detection Taskgroup uses a scoring system, from zero to ten, to rank the likely impact of any public announcement about the discovery of an extraterrestrial intelligence (ETI). The **Rio Scale** was first proposed in October 2000 by Iván Almár of the Konkoly Observatory, Budapest, and Jill Tarter of the SETI Institute, during the fifty-first International Astronautical Congress in Rio de Janeiro. Their paper was entitled "The Discovery of ETI as a High-Consequence, Low-Probability Event".

This question of impact versus likelihood is a balance that responsible governments must consider all the time when planning for possible catastrophes. Viral outbreaks and terrorist attacks, for instance, are considered high-consequence, high-probability risks, so a good deal of cash goes into trying to prevent them. The probability of an asteroid smashing into a city and destroying it is deemed very low. Nevertheless, many governments fund at least a modest amount of research into asteroids because the consequences of any such smash would be very great indeed.

The Torino Scale is a deadly serious analytical tool for assessing asteroid threats. Small asteroids and bits of space dust fall on the Earth every day. The big crater-makers hit, on average, once in 10,000 years, and city-sized destruction occurs perhaps once in 100,000 years. But that doesn't mean we have no need to worry about asteroids for thousands of years to come. On the contrary, some calculations suggest that we're long overdue a significant hit. Detecting dangerous asteroids and other "near-Earth objects" (or NEOs) is now a priority for NASA and other space agencies. Modest-sized rocks frequently drift between the orbits of Earth and the moon, and in recent years, one or two haven't been at all modest.

We can't claim that the detection of an alien signal would necessarily be as tragic as an asteroid impact. Fingers crossed, an ETI detection would be a positive event rather than a destructive one. Nevertheless, there are legitimate issues of concern. Accordingly, Tarter and Almár stated that the news of a discovery "would be like the announcement of the impending impact of a large asteroid – another example of a potentially high-consequence, low-probability event". Mathematically, they phrased their Rio Scale in a similar way to the Torino Scale, so that government risk analysts could think seriously about the social disruption that might be generated by confirmation of an alien signal. The Rio Scale, its authors propose, "may be our best chance of avoiding misinterpretation and sensationalism".

The Rio Scale balances three significant factors: the credibility of a signal detection; the source's proximity to Earth; and whether the signal is omnidirectional and aimed at no one in particular, or a focused beam directed specifically towards Earth. Imagine these three possible scenarios:

▶ **A signal detected from 50,000 light years away** seems artificial, but we cannot perceive any specific information content, and it may just be accidental leakage of noise from an alien society. We have no way of knowing if its senders are still around after a radio time lag of five hundred centuries. The general case for the existence of at least one alien intelligence is proven, but other than reshaping our philosophies, it has little immediate impact on day-to-day human affairs, because we have no sensible prospect of sending any message back, let alone establishing a dialogue. That's a relatively low score on the Rio Scale.

▶ **An omnidirectional signal detected a hundred light years away** causes ripples of consternation, as well as excitement. In theory at least, we could reply within the next human generation, and let the aliens know we've found their beacon. This is a notch higher up the scale.

▶ **A specific, intelligible signal from a few tens of light years away** is detected, aimed directly at us. This is the scenario that would have all the world's governments in a serious panic. Should we risk replying to intelligent entities so close, in terms of interstellar distances, that we could converse back and forth within our lifetimes? On the other hand, if we failed to reply, would we be missing out on humanity's greatest opportunity for scientific, philosophical and even spiritual progress?

Much depends on whether we think any aliens we encounter would be friendly or hostile (Hollywood has given us many reasons to fear the latter). This debate has exercised us for decades, and the only way we can explore it is by looking into the mirror and contemplating our own reflections.

Friends or foes?

If we did make contact with an alien race, could we be sure of their friendly intent? Tracing the history of initial encounters on our own planet might give us some clues. Looking down the pessimistic end of the telescope, we can imagine a Polynesian islander standing on the shores of her little universe a couple of hundred years ago, watching huge and complex wooden ships appear on the horizon, their white sails billowing majestically in the breeze. That islander might have thought, "How god-like, how far beyond petty spear-throwing and tribal strife the owners of such vessels must be! There can't possibly be anything they need from us and our tiny little island! I'll give them a pretty necklace of seashells as a welcoming gift…"

Hawking's warning

Professor Stephen Hawking, renowned physicist, ardent space enthusiast and author of *A Brief History of Time* (1988), suggested in 2010 that aliens could simply raid Earth for its resources and then move on. "We only have to look at ourselves to see how intelligent life might develop into something we wouldn't want to meet." He drew a bleak portrait of a nomadic culture living in massive ships, moving from planet to planet like spacefaring locusts, consuming whatever they find, having used up all the resources on their home world. "If aliens ever visit us, I think the outcome would be much as when Christopher Columbus first landed in America, which didn't turn out very well for the Native Americans." While Hawking finds the possibility of extraterrestrial intelligences scientifically fascinating, he thinks that any attempt to communicate with them is "a little too risky".

Despite winning a significant victory in the US/USSR space race, Apollo 11 didn't pave the way towards a genuinely advanced spacefaring culture.

Our history is proof that you never know what strangers might want from you. The Aztec people of sixteenth-century Central Mexico prized beautiful bird feathers far above gold, and were dumbfounded by the Conquistadors' obsession with the stuff. History also tells us that most contacts between a technologically advanced civilization and a less advanced society end in the exploitation of one by the other. No prizes for guessing which way round that process works.

In similarly gloomy vein, if someone unfamiliar with the global politics of 1969 had watched, awestruck, as **Apollo 11**'s giant Saturn V rocket lifted off for the moon, they might well have concluded that our great and noble species was heading for celestial nirvana. A glance at the other stories in that week's newspapers (Vietnam, rioting in Honduras, a crazed killer on the loose in California) would have revealed a sadder truth. Apollo 11 was an uplifting interlude in a decade of war and social unrest, but it wasn't a genuine step on the road to the stars. Saturn V and its ingenious lunar payloads were spurred by Cold War competition that threatened the destruction of all life on Earth. Technological advancement doesn't guarantee peacefulness. On the contrary, it enables more powerful weaponry.

Just how civilized must a culture be before it relinquishes the desire to dominate less developed nations? Citizens of Europe and North America no doubt regard themselves as belonging to the most liberal and advanced societies on Earth, but Third World nations are still suffering the consequences of generations of exploitation. The relationship between poor and rich nations is a complex and dynamic brew: even well-intentioned interventions don't necessarily end in harmony. The current global perspective of many diplomats and politicians enables the "exploiters" to recognize the value of helping, rather than hindering, the "exploited". Environmentalists, for instance, know that if the rainforests of South America are to be protected for the benefit of the planet as a whole, then regional economic problems have to be addressed. But how do you reconcile the needs of global society with the rights of rainforest communities, whose survival may depend on clearing land for farming, or hunting animals that have sentimental value only for outsiders with little knowledge of local life?

When it comes to SETI, the most frightening possibility, of course, is a hostile alien invasion of the Earth, as depicted in countless science-fiction movies and TV shows. There are compelling (and, we must hope, valid) arguments against this scenario. Arthur C. Clarke believed that technologically advanced alien civilizations will have solved all their energy problems by the time they embark on interstellar voyages of discovery. He reasoned that societies with limitless resources would also have solved most domestic problems on their home worlds. Warfare and social struggles are usually a product of unfair wealth distribution, and wealth is simply the monetary metaphor for resources, which, in turn, all stem from energy.

The ability to travel between far-flung worlds might indicate a civilization at peace with itself. With boundless supplies of energy on tap, it might be able to adapt barren worlds for habitation and spread its teeming populations to new resource-rich territories. Such a culture should be capable of dispensing with close-fought internal competition altogether. If its explorers ever visited Earth, they would have nothing to gain from exploiting us. No amount of pillage and destruction of this one little planet could match what they could already derive from the broader galaxy, or so Clarke reasoned.

Incidental destruction

Leaving aside the question of deliberate exploitation, interplanetary visitors may bruise us unintentionally. We believe dolphins to be almost as intelligent as humans, yet, in some parts of the world, fishermen snare dolphins in their nets while pulling in shoals of tuna. Capturing a profitable tonnage of tuna fish is considered more important than an individual dolphin's life. A variation on this theme applies to primate medical research. We acknowledge our kinship with chimpanzees, and regard them as sentient beings worthy of our protection and affection. At the same time, we subject them to painful surgical and psychological experiments precisely because of their similarity to us. The author of this book eats some animals, but keeps others as pets that he wouldn't dream of eating. Might alien visitors display similarly muddled attitudes towards humans? Could we become the victims of incidental, collateral damage in their grander schemes for mining whatever wealth or resources they can from the universe? (In Douglas Adams' *The Hitchhiker's Guide to the Galaxy*, Earth and everything on it is casually destroyed to make way for a hyperspace bypass that, presumably, benefits a wider galactic community.) Even when we think of ourselves as responsible nature-lovers with a deep respect for other living things, each innocent step we take on the path to the donkey sanctuary crushes an insect.

The Kardashev Scale

Most societies have dreamed of a world without hunger or strife: one where life is better, where milk and honey are always on tap, and no one knows fear or depression. Over the last two or three centuries, rationalists have wondered how we can reach this goal through clever management and subtle machinery. At the dawn of the twenty-first century, we doubt whether any amount of organizational brilliance can tame the world's chaos. What if that pessimism is misplaced – at least for some alien races?

Astrophysicist and pioneering Russian SETI researcher **Nikolai Kardashev** devised the Kardashev Scale, a speculative ranking of alien advancement on the basis of energy access. He published his scheme in a 1964 edition of the *Journal of Soviet Astronomy*, and it's been a spur to SETI thinking ever since. He proposed three levels of alien civilization:

▶ **Type I** makes the fullest and most efficient use of energy on its home world.

▶ **Type II** harnesses the power of its sun to the fullest possible extent.

▶ **Type III** gains access to the effectively limitless resources of its galaxy.

Freeman Dyson, a revered elder statesman of modern physics, thinks that humanity should be capable of reaching Type 1 status in a couple of centuries, so long as we work at it. Kardashev thinks that Type II ranking would take another thousand years: a long time in the scale of individual human lifetimes, but the merest blink of an eye in evolutionary terms.

Even now, despite our seemingly intractable crises of warfare and environmental hazard, optimists sense the beginnings of a Type I civilization here on Earth. Arthur C. Clarke predicted the coming of a "global village". English is a commonly accepted global language, the Internet is a global communication system, and we now have a global economy and global culture. Many people mourn the swamping of individual cultures amid the irrepressible buzz of Anglo-American-inspired consumerism, although the global village is evolving to take in Asian, South American, Indian and countless other once-localized influences. Terrestrial civilization shifts and shimmers, cracks and mends, and the unifying glue is technology, the ever-more efficient transportation of people, materials, foodstuffs and, above all, ideas.

The greatest threat to the emergence of a Type I global village is that we are forced to compete, on a petty local scale, for natural resources: land, water, fishing rights, minerals, oil and gas. Limited resources create strife. For the time being at least, our mediocre energy exploitation qualifies us for **Type 0** status. We derive almost all of our energy in primitive ways: by the inefficient combustion of fossil fuels and clumsy methods of second-hand heat extraction from lumps of uranium and plutonium coupled with steam turbines, even as a vast source of energy beams down on us for free. The sun emits energy in all directions. Earth intercepts about one billionth of the total output. Humans harness a minuscule fraction of that one billionth share. Not very clever. It's like trying to nourish ourselves by eating one grain of rice from a sack and discarding all the rest, because we don't know how to make a bigger hole in the sack – or how to reach anything other than the bottom row of sacks in a colossal warehouse piled high with them...

If we could free ourselves of energy shortages, we'd become a Type I society – although one still trying to survive in a dangerously constrained environment on the surface of a small ball of rock, under constant threat of annihilation from viral pestilence or environmental carelessness. Even a Type I civilization is fragile. Stephen Hawking has said repeatedly that our long-term survival depends upon our expanding into the solar system, just in case anything drastic occurs to the one planet on which we depend at the moment. Shit, as they say, happens. Throughout history, we've often sensed disaster coming from just around the corner. It usually comes, though not always from the corners we're worrying about at any given moment, nor at the time we expect it.

Type I and Type II doomsday scenarios

An Assyrian clay tablet dating from 3000 years before the birth of Christ bears the words: "Our Earth is degenerate in these latter days. There are signs that the world is speedily coming to an end." Pope Innocent III expected Armageddon to take place in the year 1284, exactly 666 years after the rise of Islam. The thirteenth-century Czech doomsday specialist Martinek Hausha warned that the world would be dust by February 1420. An industrious succession of priests, soothsayers, charlatans and opportunists have predicted our end with gleeful enthusiasm. Modern scientists have their doomsday scenarios too, based on better theories than those of ancient astrologers and mystics, yet still flavoured with guesswork and speculation.

The millennium celebrations for the year 2000 were haunted by the fear of a global meltdown caused by date-sensitive computer chips. Geologists

Geologists believe that falling water levels in Yellowstone National Park may be a sign of an impending eruption that could blot out the sun for decades.

have warned that Vesuvius, the volcano that destroyed Pompeii in AD 79, is about to strike again, threatening the lives of millions of people in the Bay of Naples; and **Yellowstone National Park** in the US is apparently ready to blow. Park rangers know that water levels in the local lakes seem to be falling. As the shorelines slowly dry up, it takes that little bit longer each year to walk from dry ground to the water's edge. Yet the volume of water remains the same, as does its depth. Geologists have made sense of the conflicting data. It isn't the water dropping, so much as the surrounding landscape rising, as a vast underlying blister of magma pushes up against it. If the crust breaks and the magma erupts, the resulting dust cloud will be so huge, it will blot out the sun for decades. This cataclysm could happen within the next century, or in a hundred thousand years. It could ruin your lunch next Thursday. Nobody knows.

Climate change experts predict wide-scale flooding of coastal regions in the next generation or two. Other climatologists believe that the increased cloud cover caused by global warming will work a more subtle disaster. So much sea water will be evaporated in the heat, the entire planet will be girthed in cloud all year round. Temperatures will then fall, generating a new ice age. To cap it all, asteroid-watching space agencies have calculated that a major city-destroying meteorite should slam into us once every 100,000 years, and an "extinction level" impact should occur once every 50 million years. We may be overdue for both. Oh, and did we mention? The weapons from the Cold War haven't disappeared. They're merely awaiting some unexpected resurgence of thermonuclear tension.

Okay, so this human circumstance of threat and peril may not be typical for all galactic civilizations, but consider the implications of the Mediocrity Principle, the compelling argument that the laws of nature must be the same throughout the observable universe. Stretching this idea a little further, it's not unreasonable to assume that any technological alien society will learn to harness the same physical and chemical forces that we've discovered. First will come fire and heat, and other fairly basic conversions of energy. Then, as the secrets of matter are unveiled, all the possibilities of nuclear energy will be unleashed, with their attendant risks of accident or planet-scorching warfare. How can any civilization, alien or human, ensure its long-term survival in the face of such dangers? Expansion across space seems a compelling possibility – not just journeying through it, but reorganizing it at colossal, solar-system-spanning scales.

Dyson Spheres

In 1959, physicist **Freeman Dyson** of the Institute of Advanced Study in Princeton, New Jersey, wrote a paper for *Science* entitled "Search for Stellar Sources of Infrared Radiation". He suspected that intelligible alien radio signals would be extremely rare and hard to find. Instead, he wondered if we might seek out the faint infrared warmth leaking from extremely large-scale alien technologies. In his 1979 autobiography *Disturbing the Universe*, he recalls conjuring up rough engineering designs of the machinery required "to take apart a planet the size of the Earth and re-assemble it into a collection of habitable balloons orbiting around a sun." An alien civilization might construct great swarms of energy-gathering panels, assembled from the wreckage of asteroids or unwanted (and one would hope, lifeless) planets and moons. The panels would drift in a wide, spherical orbit around their solar system's sun, trapping a large proportion of its radiated energy inside a colossal shell.

From the viewpoint of our astronomers, that sun would seem not to exist, because its light and other energies would be hidden from view by the shell.

> **"Since civilizations always face problems that require continuously greater activity, it is likely that supercivilizations will undertake activities and construct structures of a very large scale."**
>
> Nikolai Kardashev, *On the Inevitability of Supercivilizations* (1984)

But what if we detect the faint infrared warmth of the shell itself? Even the most efficient energy-trapping materials must get a little warm, and leak some of that warmth into space. An extremely large, spherical source of gentle infrared heat would be hard to pass off as any kind of natural sun. Infrared signatures aren't what astronomers usually look for when seeking out bright, hot stars. Scientists searching for intelligent life in the universe should look for low-energy infrared coming from an object larger than a sun, yet smaller than a solar system, Dyson proposed. A civilization with access to such a substantial share of its sun's energy output would be Type II on the Kardashev Scale, and might be capable of supporting trillions of individuals living on the sunwards-pointing inner surfaces of a shell's many panels. This concept is known as a **Dyson Sphere**.

It's not a complete fantasy. The eminent British molecular biologist J.D. Bernal (1901–71) speculated as early as 1929 about the starter steps. He envisioned space habitats for thousands of people, known in his honour as Bernal Spheres. In the years before the loss of innocence caused by the explosion of the space shuttle Challenger in 1986, NASA gave serious thought to the possibility of orbital settlements for vast numbers of people. With the US space effort somewhat adrift right now, it's hard to imagine a time when Congress listened, in rapt attention, to a charismatic lecturer proposing vast artificial worlds in space as a way of easing environmental pressures on Earth. The structures NASA envisioned, at least 3km long, would have supported tens of thousands of people, all living in comfortable, leafy suburbs. Throughout the 1970s, it was possible to talk about space colonies – essentially, precursor Dyson Spheres – without sounding like a loon.

Human colonies in space

Dr Gerard K. O'Neill (1927–92), physics professor at Princeton University, was renowned for his work on particle colliders, but by the end of the 1960s, he was on the lookout for new inspiration. He found it in the wake of Apollo 11's lunar touchdown. As he recalled for a NASA interview a few years later, "It just seemed to me that to be alive at that time, and not to try to take part in that unique event in human history, the first breakout from the planetary surface, would be something I would regret forever."

At first, O'Neill incorporated space exploration just as a theoretical concept to stretch the imaginations of his students. Then something remarkable happened. A small group of students took up O'Neill's challenge with enthusiasm, and together they and their tutor formulated a plan for space that addressed two of the major concerns of people on Earth: the 1970s energy crisis, triggered by international oil embargoes, and the growing awareness that the climate was in trouble. One of their primary inspirations was *World3*, a computer simulation that investigated the links between population, food production, resources and industry, and the effect these had on the planet.

> **"Given plenty of time, there are few limits to what a technological society can do. In one man's lifetime we cannot go very far. But a long-lived society will not be limited by a human lifetime."**
>
> Freeman Dyson, *Disturbing the Universe* (1979)

The simulation was the work of the influential consultancy group known as the Club of Rome, founded in 1968 by Italian industrialist Aurelio Peccei and British scientist Alexander King. The group's purpose was to help solve perceived global problems. In need of disciplined numbers and data about the environment, they commissioned a team of analysts at MIT, led by biochemist Donella Meadows. The resulting book,

The Limits to Growth (1972), was an international sensation. It stated that the Earth was running out of resources, and that humanity's ever-growing population could not be supported indefinitely.

Spurred by this, O'Neill championed the idea of "space as a place for people to be in, not just to look at". In January 1976, he testified before the US Senate Subcommittee on Aerospace Technology and National Needs. "The long-term survival of humanity, and of the plant and animal life that we cherish on Earth, can best be assured by building colonies dispersed throughout our solar system and beyond", he urged. A large-scale human presence in space could "reduce the constant threat of wars by opening a new frontier with virtually unlimited new lands and new wealth."

O'Neill's technical research hit a nerve with its detailed studies of finite (closed-loop) ecological systems, an urgent concern for environmental campaigners. Everything inside his space colonies would have to be recycled, from food packaging to human waste. Especially popular was his proposal that all heavy industry and energy production should be taken into space, thus saving Earth from the burdens of pollution. His book, *The High Frontier: Human Colonies in Space* (1976), became a bestseller. It highlighted three colony designs:

▶ **Island One** A rotating sphere, similar to a Bernal Sphere, more than a kilometre in circumference, with 10,000 inhabitants occupying the equatorial region.

▶ **Island Two (Stanford Torus)** A doughnut-shaped world capable of supporting 140,000 people. A free-flying mirror floating near the settlement reflects sunlight into the living areas of the torus.

▶ **Island Three (O'Neill Cylinders)** The ultimate design. A pair of cylinders, each 32km long and 6km in diameter, are linked at their ends by adjustable beams, and each rotates in the opposite direction to the other: a simple way of maintaining their correct orientation to the sun gyroscopically, without expending thruster fuel. Inside the cylinders, three lengthways land areas are separated by three strips of windows, illuminated with mirrors that open and close to create a familiar cycle of day and night. Each cylinder rotates about forty times an hour, simulating Earth's gravity. A total land area of 800km^2 accommodates several million people. An outer ring of agricultural greenhouse pods provides food.

First, a return to the moon was required. A lunar mass driver, a sort of electric rail gun, would be built on the moon's surface to hurl raw materials into space for the colonies' construction. O'Neill reckoned that the mass

Doughnuts in space: NASA artwork of a toroidal colony, housing around 10,000 people, painted in the heady years after the first moon landing.

driver would call for "a year's worth of shuttle flights". The problem was that everyone in the 1970s expected too much from the forthcoming space shuttle, which had yet to make its first flight. No one among O'Neill's team gave serious thought to the problem of launch vehicles, let alone the differences between an Earth-orbiting shuttle and the hardware required for the first post-Apollo touchdowns on the moon. Niggling details about accommodating the first teams of mass driver engineers and orbiting construction crews were all brushed aside. "We would use the shuttle's external tanks to make modular living-quarters for use in low and high orbit and on the lunar surface", O'Neill wrote, somewhat optimistically.

It didn't happen, because of the somewhat catch-22 circumstance that if we want to get better at harnessing the energies of space, first we have to get better at harnessing the energies needed to get up there in the first place. Despite the difficulties, a new generation of space entrepreneurs is making steady progress towards building small space stations and orbiting hotels, as monolithic NASA – embattled by funding shortfalls and political stasis – cedes ground to a plethora of small, private companies with innovative ideas about how to play the space game. Watch, as they say, this space.

Of interest to those SETI theorists expecting an alien expansion into our neighbourhood is the fact that O'Neill never once considered colonizing other worlds. "Mars, or any other alternative planetary surfaces, are fairly unpleasant options. They are the wrong distance from the sun, and have the wrong rotation times and wrong gravities." He saw no sense in giving up the massive amounts of solar energy available in near-Earth space in favour of the cold, dwindling sunlight available on a distant planet. By the time we become a Type II civilization, we might choose to get by with a Dyson Sphere around our own sun, and go no further. And perhaps other intelligent civilizations are already choosing to do likewise?

But, hang on a minute, 5000 million years from now, a Dyson Sphere will be of no further benefit to us. The sun's energies will fade, leaving us (or more likely, our unimaginably strange descendants or evolutionary replacements) cold and doomed. As a Type III civilization, we could make our presence felt not just throughout our re-engineered solar system, but across the galaxy, thereby becoming immune from localized stellar fade-outs. Theoretical physicist John Wheeler (1911–2008) coined the term **black hole** to describe the superdense core of a collapsed supernova, whose gravitational field is so strong, even light cannot escape. He wondered if we – or clever aliens – might exploit these immense forces in some way. At the very least, we could harness the power of other stars and make more Dyson Spheres. Wheeler also coined the term **wormhole** to describe the possibility of creating shortcuts through space time. This is how science-fiction writers imagine starships travelling almost instantaneously from place to place.

Wormholes are mathematical abstractions rather than feasible options for space travel. At least, they are for this generation. Professor Michio Kaku of the City University of New York works at the extremes of physics, where far-out ideas are everyday meat. "With recent advances in quantum gravity and superstring theory, there is renewed interest in energies so vast that quantum effects rip apart the fabric of space and time," he says on his entertaining and wide-ranging personal website. "Although it is by no means certain that quantum physics allows for stable wormholes, this raises the remote possibility that sufficiently advanced civilizations may be able to move via holes in space."

Perhaps, when the five hundredth edition of this book appears in some as-yet undreamed of quantum-electronic format in several centuries' time, we can speculate about wormhole-tunnelling starships. Not yet, alas.

The end of evolution?

Suppose some advanced alien beings are sufficiently wise to survive all the crises thrown at them – either by accident or their own ancient idio-

The zoo scenario

Given the current state of our physics, it's not entirely safe to assume that aliens couldn't master the galaxy via wormhole superhighways. Is the fact that none of them have come calling proof that the laws of physics as we understand them really do forbid swift interstellar travel? Well, perhaps not.

A solution to the Fermi Paradox could be that advanced alien species might be so sophisticated in their moral sensibilities that they wish to leave us alone for fear of wrecking our culture, or exerting undue influence on us lesser beings. In the television series *Star Trek*, a future generation of human space explorers obey a "Prime Directive" forbidding interference with any alien civilizations less technically developed than theirs. Of course, *Star Trek*'s weekly doses of drama often called for plenty of interference, but the concepts behind the Prime Directive were interesting. If television scriptwriters can dream up such an idea, maybe aliens can too. They may be the extraterrestrial equivalent of naturalists camouflaging themselves when studying birds and other easily alarmed creatures at close quarters.

One extreme possibility (the zoo hypothesis) suggests that Earth is protected from any visits at all, just as some especially sensitive wildlife reserves are protected from disruptive tourism. According to David Grinspoon, a space scientist from the Southwest Research Institute in Colorado: "They might not want us to know they are there. They might be protecting us, or protecting themselves from us, and waiting for the right time to make contact."

Captain Kirk and his crew do a bad job of camouflaging themselves on a visit to twentieth-century Earth in *Star Trek IV: The Journey Home* (1986).

cies – and, as a result, prosper and develop indefinitely. Arthur C. Clarke believed that a successful strain of cultural and technological evolution could have only one outcome, and that very few species would attain it. A race could become godlike, at least if witnessed from our less exalted perspective. According to one of Clarke's favourite maxims: "To the primitive mind, any sufficiently advanced technology would be indistinguishable from magic." He suggested that, in their infinite loneliness, such glorious entities might turn to little creatures like us for companionship. They would seek the kinship of one consciousness with another, the same kinship that makes us search the heavens in the hope of finding them. On the other hand, they might try to tame us as pets.

In 1968, Stanley Kubrick, director and co-writer of *2001: A Space Odyssey*, said that aliens at the top of the cleverness pyramid would seem like gods to us mere mortals. As he told an interviewer from *Playboy* magazine: "These beings would be gods to the billions of less advanced races in the universe, just as man would appear a god to an ant that somehow comprehended man's existence … and if the tendrils of their consciousness ever brushed men's minds, it is only the hand of God we could grasp as an explanation."

Godlike powers could be too much of a good thing. Imagine an alien culture freed from all needs, wants and hungers. Would it become stagnant and fall into decadence and chaos? Media commentators see signs of moral and social breakdown in supposedly advanced modern nations: in the US and the UK, for instance. Many economists are worried that while India, China and South Korea are interested in building marketable new technologies, the once-proud powerhouses of Western industrial innovation seem to be faltering. Economic and technological empires can rise and fall without bloodshed.

Philosopher of economics John Gray insists that "progress is an illusion". In his bestselling book *Straw Dogs* (2002) he warns, "Most people today think they belong to a species that can be master of its destiny. This is faith, not science … To think that science can transform the human lot is to believe in magic." Karl Popper (1902–94), arguably the greatest philosopher of science of the twentieth century, warned that automatic assumptions about social and moral advancement through science and technology are bound to be flawed. His 1945 essay "The Open Society and Its Enemies" tells us, "Our dream of heaven cannot be realized on Earth". He was dismissive of futuristic utopian visions of perfect societies whizzing between gleaming new cityscapes linked by glass tubes and jet-pack routes. "Only when we are in possession of something like a blueprint of the society at which we aim can we begin to consider the best means for its realization", he said; and that blueprint, of course, is an impossible goal, "owing to our limited experiences".

There's nothing inevitable about our climb towards a better, more advanced state of being. We might just as easily slip down the Kardashev Scale and into the sub-zero levels, like some supposedly gold-plated cosmic investment suddenly gone wrong. The Drake equation (see p.131) specifically addresses this possibility in the term "L", the length of time during which an alien civilization might be capable of communicating with us. One possibility that exercised the minds of SETI theorists during the Cold War era was the fact that the invention of nuclear power gives rise to the prospect of **planet-scale destruction**.

We have to assume that any technological alien society would encounter this hazard as soon as they learned about atoms and nuclear energy. Could they get past it and move on? So far, we humans think we've managed it. The worst crisis came in the autumn of 1962, when a US–USSR stand-off over missile deployments in Cuba and Turkey nearly led to the end of the world. Since then, the threat of nation-to-nation thermonuclear exchange has lessened somewhat, but the weapons are still out there, and could be detonated in some as-yet unforeseen crisis.

British biologist J.B.S. Haldane, the same man who speculated about a "prebiotic broth, or primordial soup", gave thought to this prospect two decades before the first nuclear reactors were conceived. While serving with the British Army in India, he attended a dance party one night in 1918. There was much talk of Nova Aquilae, a star that blazed unusually bright for a few weeks in the sky that year. With the slaughter of World War I fresh in his mind – or perhaps just seeking respite from noisy revellers – Haldane went outdoors to look at the sky. He speculated about "a too successful experiment in induced radioactivity" on a distant alien world, created by a society in which "too many men came out to look at the stars when they should have been dancing".

Must all civilizations fail?

A more subtle doom is that civilizations can simply drift apart of their own accord. "This is the way the world ends", wrote the mournful modernist poet T.S. Eliot, "Not with a bang but a whimper." In the final chapter of his renowned appraisal of Western civilization, *The Ascent of Man* (1973), scientist and historian Jacob Bronowski observed, "I am infinitely saddened to find myself suddenly surrounded in the West by a sense of terrible loss of nerve, a retreat from knowledge." He could not help but "mourn the passing of the particular civilization that nurtured me." A great deal of science fiction deals with interstellar civilizations of tremendous age. We have no model to suggest that such durability is even possible. On the contrary, everything we know about our world points to change as the only eternal truth.

To quote another great poet, W.B. Yeats, "Things fall apart; the centre cannot hold." The power of the Roman state at its height seemed limitless, but by the middle of the fourth century AD, there were signs of decay. The military support structures were weakened, the economy corrupted and the empire fell prey to the armies of supposedly inferior Germanic tribes. Such a pattern has been typical throughout history. Mighty empires become diffuse and uncontrollable. As their geographical spans increase, so the central power structures lose touch with events at the outermost limits of their reach. If you think about it, many romantic holiday trips are all about gazing reverently at the ruins of ancient civilizations who thought they were building for eternity.

Earthly empires have always been militaristic, and that will always be the problem. Other civilizations try to knock them on the head, and with that level of tension, things are perpetually on the verge of chaos. Assuming that war is no longer a factor among Type III aliens, the collapse of such a civilization wouldn't have to be sudden or brutal. A culture freed from hunger, competition and other driving forces by its own powerful technologies can disintegrate without the need for invasion by hostile forces. Creatures smart enough to design giant starships could become too lazy or atrophied to build them. They might never follow in the drone-steps of the self-replicating machines they've sent out to explore the galaxy for them (see Chapter 11). Cleverness alone is not enough to guarantee cosmic survival.

In a 1964 letter to *Scientific American*, Freeman Dyson warned that it was unwise to confuse material advances with genuine social or moral progress. "Intelligence may be nothing more than a cancer of purposeless technological exploitation, sweeping across the galaxy as irresistibly and pointlessly as it has swept across our own planet." He has also counselled, "It is easy to imagine a highly intelligent society without technology. It is easy to see around us examples of technology without intelligence."

We can speculate about all this until the cows come home. In the absence of reliable data, our hopes for a friendly, productive, uplifting encounter with ambassadors from another world hinge on faith rather than scientific evidence. We cannot predict whether an alien society will show any political or moral characteristics we might recognize. We cannot assign compassion and peaceable intentions to aliens on the basis of vague notions about what an advanced civilization might look like. We cannot assume that aliens must be nice to meet. Unconstrained beings, freed from evolutionary pressures and with all of the galaxy at their disposal, would need profound wisdom to guard against the very freedom that might destroy them – or cause them to destroy us – whether on purpose or by mistake.

The long road to the stars

Is it possible for alien explorers to visit our solar system? Could we travel into the galaxy and meet them on their home ground? New technologies promise many wonders, but there seems to be one major stumbling block that puts a limit on space travel.

The unbreakable speed limit

Einstein's laws of relativity won't permit anything solid, such as a spaceship, to travel at the speed of light. We do know how an interstellar probe might be accelerated, slowly and gradually over many months, to a velocity of over 15,000km per second, some five percent of light speed. Given current technology, a space probe would take two million years to reach the far side of the galaxy.

Even if we could build starships capable of travelling at close to the speed of light, the effects of **relativity** would cause complications for space voyagers. We think of space as having three dimensions: up and down, left and right, forwards and backwards. Objects such as suns and planets, cats and dogs, occupy particular volumes of three-dimensional space. However, in Einstein's description of the universe, a fourth dimension has to be taken into consideration. *When* does an object occupy space? Relativity wraps space and time together into four-dimensional **space-time**. Time can be stretched out or compressed just like any other dimensional measure, such as length or width. The speed of light stays constant while space-time bends and stretches around it.

Imagine a clock that counts pulses of light bounced between two mirrors inside a jet plane flying very fast, but at a constant velocity. The jet's motion is undetectable to its pilot, for the same reason that people don't spill their drinks on airliners hurtling across the sky. The light bounces back and forth between the mirrors, and the pilot notices nothing unusual. However, from the point of view of scientists observing from the ground, the pulses leave one mirror and travel diagonally to the next, because as each pulse begins its short journey, the opposite mirror has shifted slightly forward. The scientists see the pulses travelling on a long zigzag path, rather than the short perpendicular one observed by the pilot. Therefore, from the scientists' perspective, the pulses should take longer to complete each bounce. But they don't, because the speed of light is constant. Instead, relative to the time on the ground, time on board the plane stretches, meaning it passes more slowly than it does on the ground, accommodating the light's longer journey. Only at the end of the experiment, when the jet lands and its clock is compared with a similar ground-based clock, is the time disparity revealed. Less time has passed for the jet's clock than for its counterpart on Earth.

Now imagine that two astronauts, 35-year-old twins John and Jane, are the leading candidates for the first mission to another star system, four light years away. Jane is selected to make the round trip, and sets off in a ship that travels very close to the speed of light. Allowing a couple of years for the time taken to gradually accelerate the ship for the outward journey and to slow it down for the return trip, it will be ten years before Jane comes home, as measured by the high-quality digital watch that John wears at all times.

John stays behind, helping to run mission control. When his twin sister finally returns and hugs him in celebration, John is startled by the digital readout on his sister's watch, even though his training should have prepared him for the shock. For him, ten years have passed, while Jane's near-light-speed adventure means that, for her, less than three years have passed. While her ship was accelerating and decelerating she aged at roughly the normal rate, but during the phases of her journey when she was travelling at close to the speed of light, time aboard her ship was stretched (to put it loosely, it passed more slowly) relative to time back in mission control. Jane noticed nothing unusual about her watch during her flight. For her, time seemed to pass as normal. It "felt" like three years, both according to her senses and her watch. Now she finds that she is 38 years old, while her twin brother is 45.

Time is just one dimension of space-time. The effects of relativity also cause spatial dimensions to compress along an object's direction of travel. If Jane's spaceship could travel at the speed of light, its length along the direction of travel would shrink to zero. But no solid object can reach that speed. Relativity also shows that an object's mass increases with its velocity. So, as Jane's ship approaches the speed of light, it becomes so massive that no amount of energy available in the entire universe could boost its speed any further. If it could actually reach the speed of light, its mass would become infinite. Light photons can travel at the speed of light because they have no mass. If it were possible for you to hitch a ride on a photon, then from your point of view, no time at all would pass as you instantaneously hurtled from one side of the universe to the other. But, for external observers, your journey would take over 13 billion years, so they wouldn't keep supper warm for you.

Fantastic flights to the stars are not yet possible for human beings, although modern computer guidance systems routinely compensate for the time dilation effects experienced by digital clocks aboard fast-moving objects such as jet aircraft, rockets and space satellites relative to clocks on the ground. The discrepancies amount to mere fractions of a second, but they are real, and can cause significant navigational errors if left unadjusted.

Interstellar robot starship Daedalus, designed in 1978 to save human beings from long, dull voyages across the galaxy, never got off the drawing board.

Mounting an interstellar voyage

Many problems stand in the way of a *Star Trek*-style interstellar expedition, even at sub-light speeds. Firstly, the spacecraft would have to be very large, with gigantic fuel tanks and an expensive nuclear propulsion system. Ordinary rocket engines would barely get us out of the solar system. Secondly, human beings are not generally capable of planning their activities beyond the average lifespan of an individual: say, eighty years or so. Spaceship trips to the stars would be measured in many decades or, more likely, centuries and millennia.

It's not all doom and gloom for would-be galactic travellers, however. In 1978, the British Interplanetary Society in London published a detailed specification for **Daedalus**, an interstellar robot probe designed by aerospace engineer Alan Bond. Its mission would be to fly to Barnard's Star, the third nearest star to Earth, just under six light years away. Since Daedalus would be capable of achieving velocities of 45,000km per second, 15 percent of light speed, the journey would take less than 50 years. Daedalus would never return to Earth. Instead it would beam back radio messages upon arrival at the Barnard star system. The data would take six years to reach us. The project scientists could expect their first results just 56 years after the starship's launch.

NASA boffins reckon that in another century or so, we'll have spaceships capable of reaching a quarter of light speed, but they're likely to be very small and staffed by electronic, rather than human, minds. Remember, the closer an object comes to the speed of light, the more massive it becomes. Given our current scientific assumptions, 25 percent of light speed is probably the best that any substantial spacecraft could achieve. Beyond that, the fuel burned in even the most powerful nuclear engines would be wasted simply pushing against the inertia of the vehicle's additional mass. Tiny, lightweight probes, rather than vast multi-ton starships, may be best for the job.

> **"Our major objective was to carry out a feasibility study for a simple interstellar mission, using only present-day technology."**
>
> Alan Bond, "Project Daedalus"

Even with all these limitations in mind, a slow, heavy, crewed version of Daedalus could fly to our nearest neighbouring stars in a matter of centuries. Routine traffic among close-knit star systems could become common. For human astronauts, hundreds of years might seem too long to spend shut away inside a spaceship, staring at the walls and eating ready meals. Individuals would have to commit their lives to a mission – and the lives of their children, reared aboard the craft to complete the work their parents began. This sacrifice might be offset by building large, luxurious ships capable of accommodating substantial communities of people: lovers, families, parents and children. The descendants of those who set off on the voyage would be the ones to finally arrive at the target star system, and perhaps even to break bread with representatives of an alien civilization.

Sleeping through the journey

Astronauts on long space voyages could be placed into suspended animation, a kind of cold storage during which their breathing, heartbeat and general metabolism is reduced to the lowest possible limits consistent with the preservation of life. Today, some surgical procedures call for "induced hypothermia", during which a patient is deep-chilled to slow down tissue damage while repairs are made. But the surgeons have to move fast, before the chilling itself causes damage in other tissues. Mice and rats have been put into chemically induced suspended animation, but no one knows if humans could be sent into **hypersleep** for months or years at a stretch. For our generation at least, sleeping your way to the stars is not an option.

If suspended animation could be made to work, then from a spaceship crew's point of view, a centuries-long mission might seem to pass in a moment of deep sleep. Unfortunately, the Earth would not be held in cold storage as well. The crew would have to cope with the fact that their relatives back home would be ageing and dying at the normal rate. The starship's outward journey might represent an acceptable emotional challenge to hardened space professionals, but the homecoming could prove intolerable.

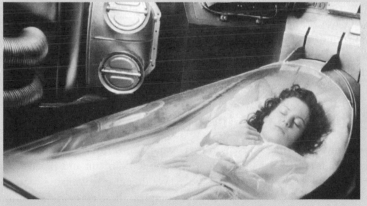

Sigourney Weaver enjoys a much-needed, age-defying hypersleep at the end of *Alien* (1979), but her loved ones continue to grow old on Earth.

Perhaps the vast time and distance involved wouldn't seem so daunting if an interstellar journey was thought of as a proportion of a given life span, rather than as a specific number of years. Consider how the mariners of previous generations were willing to tolerate sea voyages lasting many months, and sometimes several years. In the late eighteenth century, Captain James Cook and his crew made a series of exploratory voyages across the Pacific Ocean, with round trips typically lasting three years. At that time, the typical life-span of an ordinary sailor was usually shorter than 45 years. At least half of that life would have been spent at sea.

Now consider the possibilities of some future century. Medical technology will almost certainly produce significant increases in average human life-spans, at least for a privileged minority. Already, we can extend the lives of mice, and even reverse the ageing process to rejuvenate them. If humans can stay alive and healthy for two hundred years – barely doubling our current maximum – then a long starship voyage might seem acceptable to a motivated space explorer, particularly if the tedium and isolation could be eased through companionship and suitable pharmacology (sex and drugs) or suspended animation (see box on the previous page). Coming home after a century or more in space, the explorers would have a good chance of seeing their relatives again, so long as the anti-ageing treatments had been shared out fairly to home dwellers and astronauts alike.

And that's just the human perspective. It's not unreasonable to suppose that an advanced alien species might have achieved life-spans of many centuries. From their perspective, taking a few hundred years out to explore the galaxy might feel like little more than an extended vacation. But it's also not unreasonable to imagine alien beings feeling much the same way we do about undertaking long space voyages, especially if they don't find anything that interesting at the end of them. Instead, they could let their tireless – and essentially expendable – robot probes take on the discomfort of initial explorations.

Robot ambassadors

In more than fifty years of rocket exploration, human voyagers have reached no further than the moon. It's just too difficult and expensive to send people into the depths of space. We dream of colonizing Mars, or mining asteroids. Probably we can accomplish these tasks in the coming generation. Even so, this would be, as author Douglas Adams might say, "just peanuts to space".

A real avatar

Although Robonaut (see below) is capable of performing many actions on its own, its larger patterns of behaviour are driven by a human operator inside the space station, or down on the ground. A stereo vision helmet allied to a force feedback endoskeleton and a pair of gloves studded with touch-sensitive, or "haptic", sensors enables the operator to see and even feel Robonaut's environment. The system is an advanced form of what's known as **telepresence**. And if it all sounds a bit familiar, it may be because it's not a huge leap from there to the plot of the sci-fi film *Avatar* (2009), in which humans mentally "inhabit" artificial bodies so that they can explore a moon in the Alpha Centauri system and interact with the blue-skinned locals.

Interstellar voyages may be beyond us for the next several generations at least, but could be possible for our machines. It's not just the realm of science fiction: clever robots are already scattered widely across our solar system. All deep space probes and planetary landers are capable of operating semi-autonomously, so as to cope with the long time-lags between radio commands from distant Earth. Some machines, such as NASA's wheeled Mars rovers, also boast certain kinds of mechanical flexibility ("degrees of freedom"), such as pan-and-tilt camera arrays, directable radio antennae, steerable wheels, extendable soil sampling arms and other useful bits and bobs, all controlled only in the most general sense from Earth. On-board electronic intelligences have to cope with moment-by-moment decisions.

The public sometimes struggles with the idea that these clunky boxes are definable as robots, but no one will be in any doubt about one of NASA's most recent non-human space explorers: **Robonaut**. At first glance it looks just like an astronaut in a space suit. Its head is shaped somewhat like a motorcycle helmet, and its torso and arms, tipped by elegant hands with slender fingers and opposable thumbs, are very like our own. Even at second glance, it seems at least as lively as a *Star Wars* character in an armoured mask. Boba Fett springs to mind, although a ruthless bounty hunter may not be everyone's idea of a comforting comparison.

Robonaut, the first humanoid robot ever sent into space, was delivered to the International Space Station (ISS) in February 2011, tucked discreetly into space shuttle Discovery's cargo manifest. Developed by NASA in partnership with the General Motors car company, Robonaut's mission is to operate outside the pressurized shell of the station, assisting human space walkers in assembly and maintenance tasks. General Motors reckon that Robonaut-style technology will have far-reaching implications on the ground as well as in orbit, and want to see "advanced robots working together in harmony with people". NASA expects Robonaut to become a helpmate for astronauts. The space agency is also seeking a new, semi-intelligent entity that can explore the moon and Mars as though it were human – yet without the worry of feeding it, providing comfortable living quarters or arranging its ride home.

Project M: deep-space Robonaut

In November 2010, a study group at NASA's Johnson Space Center in Houston wrote the brief for **Project M**, a proposed mission to land a Robonaut-style explorer on the moon in a thousand days flat, measured from the time of budgetary approval to the moment of touchdown. (M is the Roman numeral for 1000.) The delivery vehicle would be a small, single-engined lander built by Armadillo Aerospace in Texas: one of a dozen companies making rapid advances in the private rocket business. Armadillo successfully tested a prototype lander fuelled by liquid methane and liquid oxygen. It took off from its (terrestrial) launch platform, hovered at a preassigned altitude and performed a precision autonomous landing.

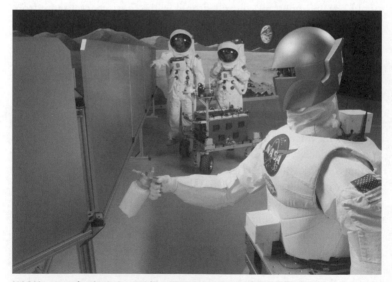

NASA's new robotic assistant does some practice welding on Earth before being shipped off to help the astronauts on the International Space Station.

Project M calls for a lunar version of Robonaut to operate in near real-time, once the Earthbound operators have acclimatized their working habits to the couple of seconds of radio time-lag in the command and feedback signals. Fanciful artwork shows the robot walking on the moon on two legs. Even if the mission does eventually get funded, bipedal locomotion is somewhat ambitious. Instead, the robot's upper torso will probably sit on top of a small wheeled platform. Further afield, real-time operations on Mars or on asteroids would most likely be ruled out because of the much longer radio time-lags. A future generation of super-smart Robonauts could operate autonomously, making their own decisions about interesting scientific targets, and sharing their discoveries with people back on Earth via delayed "telepresence" (see box on p.201).

Most robots for industrial or scientific applications look nothing like humans. Robonaut is shaped the way he is for a reason. When we become telepresent on other worlds, it's useful for our robotic representation, or avatar, to have a humanoid shape because physical and tactile information can be relayed to operators back on Earth so that they can see, feel and even hear what it would be like to be there in person. An eight-legged spider-crawler or a six-wheeled rover would sense the terrain very differently. The current generation of Mars rovers are on wheels, rather than legs, but their stereoscopic cameras are the same height above the ground, and set the same distance apart, as a pair of human eyes.

A pioneer in human–robot interaction, **Dr Cynthia Breazeal** directs the Personal Robots Group at MIT, one of the world's leading investigators of socially interactive technology. An interest in space exploration initially led her to work on systems designed for alien terrains and dangerous environments: machines with wheels or methods of locomotion inspired by the animal world. But her research soon shifted to the greater challenge of designing robots that could operate in the safe and familiar environment of our own homes. In Breazeal's own words: "Robots have been into the deepest oceans. They've been to Mars. They're just now starting to come into your home. You could think of your living room as their Final Frontier. It's really about bringing robots into our environment, rather than the other way around. Our world is constructed for our morphology: the fact that we walk on two legs, have two arms, and so forth." These developments in human technology point to an interesting possibility. If we ever encountered an alien machine, we might get some tentative clues about the physique of its original designers.

Self-replicating machines

Even as we stretch our imaginations far beyond the clouds and into the depths of the galaxy, the limit to our exploration – whether we're sending human astronauts or robotic ones – remains the sheer amount of time it would take to get… anywhere. If we really want to explore the universe one day, perhaps we'd need to put even more trust in our robotic representatives than we do at present, and let them go it alone. This is where we come to the disturbing idea of machine evolution, and self-replicating von Neumann probes…

In the late 1940s, Hungarian-American mathematician and computer pioneer **John von Neumann** (1903–57) came up with the idea of a "universal constructor", a machine capable of making copies of itself. The details were published in his book, *Theory of Self-Reproducing Automata* (1966), completed after his death by co-author Arthur W. Burks. A universal constructor would need an energy source, a computer control system and physical mobility, in the form of wheels, arms and hand-like manipulators, to gather raw materials. In von Neumann's time, universal constructors were just a theory, a logical potential. Something similar was proposed by mathematician Edward Moore (1925–2003). His 1956 article for *Scientific American* proposed the creation of "artificial living plants", floating factories – industrial plants rather than the leafy kind – that could create copies of themselves. These factories could be programmed to perform some useful task, such as extracting fresh water or useful minerals from salty ocean water. Once a few dozen of them were unleashed, they would take care of everything else, replicating to widen their scope of operations, and without further human intervention.

Today it's becoming possible to make such things for real. In 2008, Dr Adrian Bowyer from the University of Bath demonstrated a contraption that looked rather like a homemade desktop printer. Yet this tinker toy could be the technological germ that seeds a new machine species. Almost all the components in his **replicating rapid prototyper**, or "RepRap", were constructed by an earlier version of the same machine.

RepRap was derived from rapid prototyping (RP) printer technology, which converts virtual 3D computer models into physical objects. A typical RP printer works in much the same way as a conventional paper printer delivering 2D sheets of output. The extra third dimension comes from stacking successive sheets on top of each other like the stepped levels of an Egyptian pyramid. Stand back far enough, or use sufficiently fine sheets, and you don't see the jagged steps. Instead of ink or toner, RP printers use plastic powders, organic compounds or microscopic beads of metal. Other 3D-printing techniques involve firing computer-steered laser beams into tanks of translucent resins. Objects solidify at the points where the lasers intersect, like ghosts suddenly becoming solid. Most professional RP machines can output components of startling complexity, such as racing car engine components (the F1 circuit was quick to adopt this technology). Many RP printouts are useable in products designed to withstand heavy physical stresses. It's even possible to output objects with internal moving parts.

As the technology develops, becoming cheaper and faster, universal constructors could democratize and diversify the manufacturing process, opening it up to individuals and developing nations who wouldn't need vast factory production lines. Bowyer certainly sees his RepRap technology as a step in that direction: "What if you could manufacture a multitude of different things, and for a very small initial investment? They could be a powerful mechanism for elevating people from poverty." All very interesting, and potentially world-changing, but what has it got to do with exploring space and searching for life on other worlds? Von Neumann wasn't just thinking about the possibilities here on Earth. He envisaged fleets of intelligent, self-operating and self-replicating machines spreading across the galaxy on behalf of their original designers.

It may look like a fairly ordinary piece of kit, but the RepRap 3D printer could revolutionize manufacturing, and be a step towards self-replicating machines.

Von Neumann probes

The first alien envoys to Earth might not be the creatures themselves, but their advanced and long-independent von Neumann probes. They could assist a spaceflight-shy advanced civilization in taking its first steps across the galaxy, and save them from those long, boring interstellar voyages we discussed earlier. According to Michio Kaku, professor of theoretical physics at the City University of New York, a von Neumann probe is "a robot designed to reach distant star systems and create factories which will reproduce copies of themselves by the thousands".

Kaku raises the frightening conjecture of millions of probes, in a spherical swarm, expanding in all directions from their originating world. Whenever any of them encounter a useful planetary body or moon, they settle for a while, have a look around, then make more copies of themselves and send those replicants out in all directions, and so on. If the initial swarm explodes from somewhere within the heart of a galaxy like ours, 100,000 light years across, the entire thing could be investigated within half a million years, assuming a Daedalus-style maximum velocity of around 15 percent of light speed. It's a long-term strategy, though. The original alien inventors of the probes would still have to wait long centuries, or even millennia, for any of them to report back useful findings (assuming they don't become so self-aware that they decide they've got better things to do than send messages back to their inferior biological creators).

It's conceivable that von Neumann probes are already here, but are just too small for us to notice. They could be **nanobots**. According to physicist and SETI theorist Paul Davies, if this is the case, the tiny probes would be "so inconspicuous that it's no surprise if we haven't come across one. It's not the sort of thing that you're going to trip over in your back yard … if other civilizations have gone this route, then we could be surrounded by surveillance devices."

Manufacturing – say, of a fine clockwork watch – usually involves taking large chunks of metal and whittling them down into the required cogs and springs. It's an inefficient procedure, because most of the metal ends up on the workshop floor as discarded shavings. In addition, huge amounts of energy are needed to extract pure metal from crude ores in the first place, and then to melt the metal into bars, plates or rods that can be more finely cut and shaped by the watchmakers. Nanotechnology turns all this processing on its head, building upwards from the smallest scales by molecular accretion. Biology functions like this all the time. We are entering an age when the distinctions between biology and engineering are starting to blur. We will grow our machines, rather than build them. We might perhaps become capable of reshaping our world atom by atom, even though we're still not quite sure what atoms are.

If these are the trends in today's human technologies, we can apply the Mediocrity Principle (see p.29) and speculate that nothing in the laws of physics prevents a more advanced alien civilization from developing von Neumann probes on a nano-scale, just as we could. A first-generation swarm of submicroscopic, extremely lightweight space-faring devices could easily be accelerated to significant fractions of light speed without damaging their physical structures.

The Fermi Paradox revisited

The fact that we haven't found any evidence of aliens, or their supposed swarms of von Neumann probes, is sufficient for plenty of people to think we're alone in the universe – or at least, in this galaxy. Two prominent naysayers are Frank Tipler, a mathematical physicist and cosmologist at Tulane University, Louisiana, and John Barrow, an English cosmologist, theoretical physicist and mathematician at the University of Cambridge. In various publications, written jointly and individually, they weave modern technological ideas into an Earth-centred framework. An intelligence of the far future, human in its ancestry, yet no longer biological, will become godlike in its powers. Tipler proposes the total retrieval of information from the past, enabling the dead to be reborn… and amidst all this wonderment, it's likely to be us and our robotic super-descendants who will reign supreme, because extraterrestrials don't exist.

Tipler thinks that if aliens did exist, they would use their von Neumann-style **self-reproducing robot spaceships** to explore the galaxy and to contact other civilizations. The fact that we haven't seen any of these robots shows that the Fermi Paradox isn't a paradox after all. It's a clear indication that aliens aren't out there. "I believe ours is the only intelligence to have arisen in the galaxy," he says. As for the Drake equation (see p.131), he believes it already has an answer: N=1, that "1" being the Earth.

Martin Rees, Britain's astronomer royal, undercuts these screeds of pessimism with a simple statement: "Absence of evidence is not the same as evidence of absence." Carl Sagan didn't think that the absence of evidence for von Neumann probes in our neck of the woods was of any significance. Aliens of sufficient moral character and technological advancement "would have much more exciting and fulfilling objectives than strip-mining or colonizing every planet in sight." Even though we don't have the slightest shred of scientific evidence to back up Sagan's optimism, it's a nice sentiment. Surely civilized creatures wouldn't exploit planets in such a callous way? After all, we don't find strip-mining on Earth, do we?

Perhaps von Neumann probes don't have to be so exploitative. According to nanotechnologist Robert Freitas of the Institute for Molecular Manufacturing at Palo Alto, California, "a good case can be made that no replicating systems will be sent to our solar system at all. Unless life is extremely widespread, most star systems will be uninhabited." It would make more sense, both scientifically and morally, for aliens to establish "self-replicating probe factories in obviously uninhabited star systems and just to send non-reproducing exploratory probes to the fewer more promising systems."

Machine evolution

John von Neumann himself wondered if something was missing from his ideas about self-replicating machines: namely, the unconscious creative power of evolutionary pressures. In 1970, Cambridge scientist John Conway invented a computer game called *Life*. The rules were simple. Dark squares, or cells, are arranged in a grid to create what looks like a sheet of graph paper with occasional squares inked in. Dark cells stay alive so long as they have at least two neighbours, but die if they are hemmed in by four. If conditions are just right, they will generate another "live" cell nearby. The human programmer decides on an initial pattern of cells, then the computer takes over, applying the same simple rules again and again. On each iteration of the process, the results from the last run become the starting conditions for the next.

Too few cells in the initial pattern, or cells spaced too widely apart, may lead to an empty grid at the end of the game, but something quite simple, such as a single T-shaped starting pattern of dark cells, can produce amazing results. Cells breed, live and die, filling the grid with patterns of stunning orderliness and complexity. Watch the game develop, especially on a high-speed computer, and it's hard not to imagine that something in the system is "alive".

A compelling twist to *Life* is to set the conditions you want to achieve at the end of the game: a triangular array of cells, for example, or even a picture of some kind, using the dark cells as pixels (similar to the Japanese logic puzzles known as Hanjie or nonograms). The only extra requirement is a rule that compares each state of play with the desired end result, preserving the states that edge closer to it and discarding those that don't. Genetic algorithms extending these ideas are now a major trend in software.

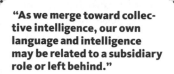

> **"As we merge toward collective intelligence, our own language and intelligence may be related to a subsidiary role or left behind."**
>
> George Dyson, *Darwin Among the Machines*, 1998

Computer scientist Peter Bentley, of University College London, suggests that genetic algorithms could supplant human intervention in the manufacturing process, requiring us simply "to nudge the universal constructors in the direction of what we want". Some of Bentley's home furnishings – produced by rapid prototyping – prove his point. "I'm probably the only person with an evolved coffee table."

The big trick with living things is that they can replicate themselves and pass on specific survival characteristics to the next generation (see p.31). The other factor, of course, is that those offspring aren't perfect copies of their parents. How could they be, when (in the case of humans, for instance) one parent's a lean, tall mum with blonde hair and the other's a squat, heavy-set dad with hair as black as night? Offspring end up with a jumble of inherited genes and minor mutations that affect their ability to survive, for better or for worse. Machines will soon exhibit some of these mix-and-match characteristics by exploiting the creative possibilities of genetic algorithms.

There's just one particular quality that distinguishes machines from the biological world: a desired end result. Unlike the design of a piece of technology aimed at satisfying some human need or desire, natural evolution isn't directed towards anything. It just happens. At the moment, machines evolve only in a default sense. The ambitions of their engineers push them towards higher speeds or better energy efficiency, boosting the fortunes of useful designs and discarding those that fail to meet expectations. We are both the inventors and the builders of machines, so they are not yet the products of a genuinely independent evolutionary history.

But the time is fast approaching when we will no longer be in the driving seat. Soon, machines will prosper – or die out – according to their own evolutionary fortunes. At some point in the future, we may encounter a disconnect between human intentions and what machines become capable of for themselves, regardless of our wishes. Machines will simply try to survive in whatever environment they happen to exist in. They will no longer be artificial. They will be just another facet of the natural "living" world. In fact, they already are. In his book *The Blind Watchmaker*, geneticist Richard Dawkins insists that machines should be "firmly treated as biological objects." They may not be alive (whatever that word means), but they couldn't arise in the absence of life.

We have to wonder if machines will evolve the power of thought. George Dyson, author of the influential 1998 book *Darwin Among the Machines* (named after an 1863 essay of the same title by Samuel Butler), is convinced that artificial intelligence "will emerge naturally, rather than as a product of some specific design on the part of human programmers". To put the problem simply, we don't yet know what consciousness is, and therefore are hard-pressed to bring it to life inside a computer. However, if we concentrate just on the characteristics we'd *like* a supposedly sentient computer to exhibit, and then unleash simple yet innumerable fragments of code into the system, something might emerge that matches our hopes – even if, when it finally comes to life, it reaches such a level of complexity that we no longer understand its intricacies.

George Dyson suggests that the **Internet** may already exhibit occasional, accidental signs of a precursor intelligence, and warns that when the first fully fledged artificial intellect does finally announce itself, "We shouldn't expect it to operate on a level that we are able to comprehend." Intelligent machines will evolve along their own path. Ten thousand years from now, they may remember humanity's role in creating them as just a brief, lucky environmental circumstance in their early history.

Biological–mechanical hybrids

Keep in mind the Mediocrity Principle as we continue our exploration of our own species and its possible future. Anything we can dream up that doesn't break the laws of physics, aliens can presumably match or outpace. In meeting the challenges of space travel, or preparing themselves for a super-sophisticated technological environment – or simply in an infuriated bid to escape the mortality of their feeble bodies – aliens might manage to rebuild themselves into something that thinks faster, works better and lasts longer. Who among us doesn't leap at the chance to counter our biological failings with the best that modern medicine has to offer?

We and our machines will become one and the same thing; there will be a union between weak, randomly evolved flesh and intentionally designed durable technologies. This may seem fantastic, even appalling, but mergers that we already take for granted, such as breast implants, artificial limbs, pacemakers and even the first electronic eyes as aids for

Are we already becoming cyborgs? Technology and flesh are perfectly combined in this prosthetic arm, which is controlled by thought commands.

the blind are paving the way towards a new phase in our evolutionary history. Perhaps we're already becoming the **cyborgs** (cybernetic organisms) much loved by science fiction. Already, few of us feel quite complete without our panoply of portable electronic devices. Soon they will be sexily wearable, and then the next phase can only be total bodily and mental integration, and a blurring of the line between us as individuals and "we" as a networked global info-species. Imagine the equivalent of Facebook in a hundred year's time and you'll have some idea of what's in store.

Alien entities may well be indistinguishable from their technologies. But we're not just talking about self-replicating spaceships or extra-terrestrial robots. On a much grander scale, they may also be indistinguishable from the planets they inhabit. We talked earlier about James Lovelock's Gaia Hypothesis (see p.49), the idea that a life-bearing planet such as Earth can be thought of as a super-organism. Lovelock's work refers just to the biological feedback of the natural environment. Suppose a massively large-scale technological alien intelligence reworks an entire planet?

US mathematician and software developer Robert Bradbury (1957–2011) imagined a series of Dyson Spheres (see p.185) built around a star, and nested like the dolls-within-dolls of a traditional Russian matryoshka toy. Some of the shells would be composed of nanoscale computers. Each shell would operate at a different temperature. The ones near the core could be almost as hot as the central star, while the outer layers would be as cold as deep space. By converting such vast energy gradients into computing power, a **matryoshka brain** could become an intelligent entity the size of a small solar system. The only limits on its vastness would be the need to keep the circuitry distances sufficiently small so that electronic pulses could be conveyed around the brain in a matter of seconds or minutes, rather than many hours. A matryoshka brain would probably think quite slowly, even if it did eventually become very, very clever.

On a more modest planetary scale, Earth is becoming a superbrain of sorts, as the Internet spreads its tendrils of copper wire, optical fibre and radio waves around the planet. If the biosphere is the thin shell wherein life on Earth exists, the infosphere is the even wider envelope that contains our data. Information and its carriers are embedded in the land, and in our houses, offices and vehicles. They fizz in the sky, snake under the sea and beam into space. As we've just mentioned, the fabric of this new entity will soon become embedded in our clothes and bodies. It's not completely fanciful to imagine a new intelligent entity emerging from a superbrain system, Web 6.0 or whatever, as it takes on a strange consciousness of its own, perhaps within the next few decades. A sentient – or at least extremely intelligent – "alien" presence might not arrive from the stars, but may well emerge right here on Earth. An intelligent being initiated by biological humans, but not biological and not human.

Science-fiction scenarios, like those depicted in the *Terminator* movies, of warfare between machines and the last of a dying breed of humans are surprisingly unimaginative. Whether on the surface of the Earth or out in the cosmos, we and our technological creations will be indistinguishable in just a very few thousand years' time. To quote Arthur C. Clarke once again, "In the future, we won't travel in spaceships. We will be spaceships." According to Seth Shostak, lead astronomer at the SETI Institute, any aliens with sufficient smarts will go down the same route. "Continuing to hunt for our equivalents – technically competent biological sentient creatures – may be an unpromising enterprise, as it focuses on a highly transient prey."

The usefulness of nonsense

When it comes to alien intelligences, countless supposedly academic papers by some of the best minds on Earth blur the boundaries between plausible ideas and speculative fancies. These theories include biological rather than electronic minds spreading across planets, vast balloon-like colonies drifting in the high atmospheres of gas giants, conscious clouds of electrically activated dust drifting through space, and so on. Carl Sagan always warned that "extraordinary claims require extraordinary evidence". Wild fantasies aren't much use for current SETI efforts, yet they do serve to remind us that aliens may be exactly that: alien. It's difficult to detect something when you don't know what signs to look out for.

It's fair to say that all of science depends on someone, somewhere, wondering, "What if...?" and then testing their ideas. Perhaps the fans of Robert Bradbury will eventually have their day. It's also fair to say that SETI's focus on primitive radio detection is a long shot at best, as is its relatively new cousin, **Optical SETI**.

In 2006, the Planetary Society unveiled the first astronomical instrument built specifically to hunt for messages in, or near, the visible light spectrum. The All-Sky Optical SETI (OSETI) telescope at Oak Ridge, Massachusetts, is operated by a team from Harvard University. They're looking for pulses of laser light. Most natural EM radiation spreads out in all directions as it travels through space. Laser radiation is artificially tweaked so that the beam remains concentrated and directional, losing as little power as possible along the way. Certainly, laser would be a valid and energy-efficient method of sending messages across interstellar distances. But we're still stuck with the same problem that we have with old-fashioned radio SETI: we have to hope that an alien civilization somewhere is using the same technology as us, and doing so in the same brief window of time that we are capable of listening for it.

The problem is that these are the tools we have, and the ones we know how to use. They are also the best techniques that the limited funding scraped together by SETI researchers allows. Best to keep things simple, as did US astronomer E.B. Frost, when he was telegraphed by a newspaper in 1909 and asked to come up with three hundred words about the likelihood of alien intelligence. In the staccato of cable-speak, he replied, "Three hundred words unnecessary. Three enough. No one knows."

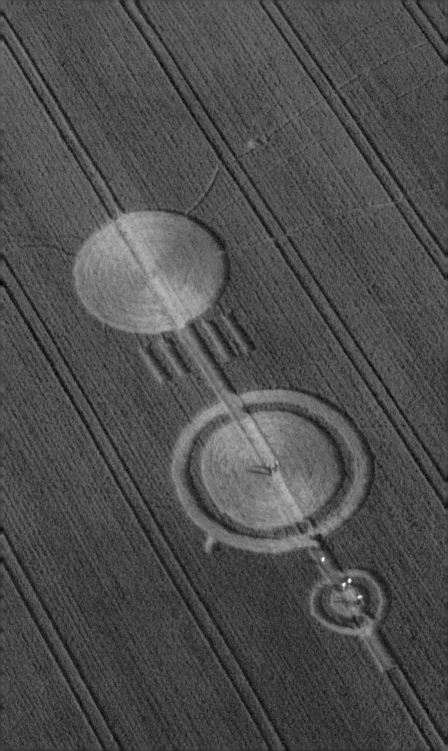

Kidnapped
by aliens

Why do thousands of people believe that extraterrestrial spacecraft have already visited us? How can we distinguish between credible "close encounter" accounts and credulous nonsense? Just because you've seen an alien in your bedroom doesn't necessarily mean you're insane.

Project Blue Book

US astronomer **Dr Josef Allen Hynek** (1910–86) specialized in the life cycles of stars and the study of binary star systems, until a new theme took centre stage in his life. In the spring of 1948, he was teaching astronomy at Ohio State University, Columbus, when three envoys arrived from the Wright-Patterson Air Force Base in nearby Dayton. They asked him about flying saucers, and he said they were stuff and nonsense. It was this healthy scepticism that won him the job as consultant for Project Sign, soon to become Project Grudge, and finally to emerge, a dark butterfly from an even darker chrysalis, as the world's most famous government UFO investigation, Project Blue Book.

The Air Force simply wanted to know what **UFOs** might be. The acronym is often misunderstood as "flying saucer" or "alien spaceship". It simply means Unidentified Flying Object. At the height of the Cold War, there was genuine concern that prototype Soviet aircraft or other unknown foreign technologies might be responsible for at least some sightings of strange lights and mysterious silhouettes in the sky. Far from investigating alien corpses in hidden bunkers, Hynek was encouraged to find natural explanations wherever possible. By and large, that's what he did, although in a 1985 interview with *Omni*, a (now defunct) US popular science magazine, he expressed unease about debunking a few of the stories he'd heard:

"The calibre of the witnesses began to trouble me. Quite a few instances were reported by military pilots, for example, and I knew them to be fairly well-trained, so this is when I first began to think that, well, maybe there's something to all this."

Among the thousands of UFO sightings he studied, certainly a handful were fascinating. One of the most famous is known as the **Michigan Swamp Gas Case** of 1966. Many witnesses observed strange lights in the skies over Michigan in March of that year, flying at incredible speeds and making exceptionally sharp turns and dives. That area of southern Michigan is pretty swampy, so Hynek was inclined to think that swamp gas phenomena were responsible. The only nagging doubt was that he couldn't be sure exactly *how* swamp gas could cause such a wide range of odd sightings. An earlier case from August 1947, initially investigated by the FBI, involved a man and his two sons witnessing a metallic object swirling down Snake River Canyon in Idaho. The object caused the top of the trees to sway. Hynek recalled, "In my attempt to find a natural explanation for it, I said that it was some sort of atmospheric eddy. Of course, I had never seen an eddy like that and had no real reason to believe that one even existed."

Dr J. Allen Hynek brandishes a photo of a UFO in 1966 – he commented that it looked more like a "chicken feeder" than a "flying saucer".

Hynek invented the term "close encounters" to categorize UFO witness accounts, and ranked them into three distinct kinds:

▶ **Close encounters of the first kind (CE1)** Mere sightings of inexplicable lights or shapes in the sky.

▶ **Close encounters of the second kind (CE2)** Apparent observations of alien craft, with associated physical effects, such as electrical interference or scorch marks on the ground.

▶ **Close encounters of the third kind (CE3)** As the Spielberg movie of the same name suggests, the third and final type of UFO encounter reaches the giddy heights of perceived interactions with alien beings.

In 1973, Hynek established the independent Center for UFO Studies (CUFOS), currently based in Chicago. Addressing a UFO symposium in Ohio that same year, he urged a strictly scientific assessment of any sightings, and wasn't afraid to waver between fascination at the possibility of alien life and distrust of the seemingly endless reports of their presence in our skies:

> "A few good sightings a year, over the world, would bolster the extra-terrestrial hypothesis. But many thousands every year? From remote regions of space? And to what purpose? To scare us by stopping cars, and disturbing animals, and puzzling us with their seemingly pointless antics?"

As for Project Blue Book, it was brought to a close in December 1969. The new secretary of the Air Force, Robert C. Seamans, Jr (fresh from a sojourn as NASA deputy administrator in the lead-up to the Apollo 11 moon landing) announced that further funding could not be justified "either on the grounds of national security or in the interests of science". According to the Air Force's figures, between 1947 and 1969, a total of 12,618 sightings were reported to Project Blue Book and its precursors, of which 701 remain unexplained. Much of the accumulated data is now in the public domain, so there's plenty of fun to be had for amateur UFOlogists.

The UK's National Archives contain a wealth of UFO-related material, also accessible to the public. A memo written in 1983 sets out the official position with admirable clarity: "The Ministry of Defence does not deny that there are strange things to see in the sky, [but] it certainly has no evidence that alien spacecraft have landed on this planet."

The Roswell incident

The most famous UFO story of modern times was widely reported in the newspapers just a few weeks before the Snake Canyon incident occurred, so we have to wonder if the one fed into the other. This was, of course, the Roswell incident.

On 8 July 1947, the *Roswell Daily Record* announced the "capture" of a "flying saucer" on a farmer's ranch near Roswell, New Mexico. The US military has insisted time and again that what was actually recovered was debris from an experimental high-altitude surveillance balloon. What we do know is that Lieutenant Walter Haut, the Roswell Army Air Field (RAAF) public information officer, issued a press release stating that personnel from the 509th Bomb Group had recovered a crashed "flying disc". The next day, a much more senior officer from the Eighth Air Force stated that a radar-tracking balloon had been recovered, and nothing more. In a subsequent press conference, some sorry-looking scraps of silvery balloon fabric were held up for inspection, and that was more or less the end of the matter for the next couple of decades (although the initial reports did trigger a wave of alleged UFO sightings in other parts of the US).

> "When you get reports from professors at MIT, engineers on balloon projects, military and commercial pilots, and air-traffic controllers, you might one day sit down and say to yourself, 'Just how long am I going to keep calling all these people crazy?'"
>
> Josef Hynek, astronomer, 1985

A succession of old-timers, publicity chancers and one or two genuinely puzzled witnesses have come forward over the years to tell their versions of the Roswell story, which has now become a vast decorative accumulation of conspiracy theories and accusations of cover-up. The US Air Force acknowledged, in 1995, that the infamous balloon was part of **Project Mogul**, a bid to peer over the horizon and detect Soviet nuclear bomb trials, using acoustic sensors and other instruments. A further official report, released in 1997, concluded that stories of recovered alien bodies may have been inspired by muddled accounts of Project High Dive, in which dummies were used to test high-altitude parachute systems.

Some conspiracy theorists point out that although Mogul was underway in 1947, High Dive didn't start until the 1950s. But how would they know? And are these conspiracy theorists actually being paid by the US government to sow further confusion? And so on, and so forth. It doesn't help that, in 1947, the Air Force was quite glad about all the reports of discs and saucers at Roswell, because it had very good reasons for not

Fragments of silvery fabric, found in Roswell, New Mexico, are identified as the remains of a crashed weather balloon by General Ramey of the Eighth Air Force.

wanting the Soviets to know what they were really doing with their balloons. We can also speculate, realistically, that newly developed bombers, missiles, jet fighters and other aeronautical objects have been spotted by witnesses unused to their strange shapes. By the time we get to see any of these sinister contraptions in the media, they've been tested in absolute secrecy for years. (Nowadays, it's usually Chinese paper lanterns that are mistaken for visiting alien spacecraft.)

Roswell and almost all other UFO sightings have prosaic explanations. Those few cases that remain completely mysterious are just that: products of our lack of explanation, not proofs of alien contact. The "U" in UFO stands for "unexplained", not "alien". It would be great if just one tiny fragment of the supposedly extraterrestrial Roswell wreckage could

Area 51

Area 51 is the popular name for part of a military base in southern Nevada, about 130km northwest of Las Vegas. Its location on the southern shore of **Groom Lake**, a flat, dry salt bed, provides a perfect touchdown surface for experimental aircraft. No one doubts that eerie new jet fighters, bombers and robotic weapons systems are tested there under conditions of extreme secrecy. One of the most beautiful aircraft in the world, the Lockheed SR-71 Blackbird, was put through its paces at the base in the early 1960s. The development programme was codenamed Oxcart, as if in ironic response to the required cruising speed, which topped three times the speed of sound. Even in retirement, nearly half a century after its first flight, the Blackbird still looks futuristic today. An over-eager observer might well be inclined to think that aliens had lent us its graceful lines. The Blackbird's original task, high-altitude surveillance, has been taken over by small space satellites.

Frequent reports of triangular UFOs in the skies around Groom Lake in the 1980s are best explained by the sinister shadow of the F-117 Nighthawk stealth fighter, another typically unusual machine from the Lockheed stable. Its existence was only acknowledged by US officials in 1988, after a decade of secret prototyping. The Nighthawk's flat, black polygonal surfaces were designed to absorb and deflect radar. When its three-point arrays of navigation lights were glimpsed at night, it certainly did look weird. Even the Nighthawk is old-fashioned now, as we enter the age of robotic unmanned aerial vehicles (UAVs). You couldn't, as they say, make this stuff up.

No matter how sleek the styling of Groom Lake's military machines, none are proof of extraterrestrial technology being hidden at Area 51, nor of warplanes retro-engineered from advanced alien counterparts that crash-landed decades ago. Secrecy is the default position of military-industrial complexes the world over. US government officials maintain a position of denial, saying, "Neither the Air Force nor the Department of Defense owns or operates any location known as 'Area 51.'" This notorious name probably has nothing to do with the Groom Lake ground facilities, and may owe its origins to the zone of highly restricted airspace above. We cannot be unduly surprised that the military wishes to discourage unauthorized overflights.

emerge from someone's attic and be submitted for serious scientific analysis. Until that happens, Roswell is just a bunch of fairy tales. Sorry for spoiling all the fun.

Alien abductions

UFO sightings are just the start when it comes to our fascination with little green men. Some people claim to have had much closer encounters. Is there a rational explanation for the bizarre fact that so many people think they've met visitors from outer space? The literature on supposed alien visitations is vast, and those with an interest in such stories can turn to many resources other than this book. Keeping to the realms of known science, it's worth looking at at least one of the various possible explanations for the widely reported phenomenon of alien abduction.

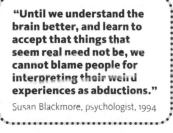

"Until we understand the brain better, and learn to accept that things that seem real need not be, we cannot blame people for interpreting their weird experiences as abductions."

Susan Blackmore, psychologist, 1994

Leaving aside the frauds and self-publicists, hundreds of perfectly sane people genuinely believe they've been visited by aliens. Just because a person believes in something, doesn't make it true. Some reports suggest that as many as three million US citizens have had such an experience, but these figures, coming from UFO enthusiasts rather than the scientific community, are unproven. Whatever the numbers involved, the interesting question is: why does the impression of being abducted by aliens feel so real to those who experience it?

In a typical account, a victim wakes in the middle of the night and finds a grey, dark eyed alien in the room. Its thin, spindly body supports a huge head. It draws its victim into a spaceship, where intrusive investigations are conducted, often of a sexual nature. When the abductee returns to the normal earthly realm, he or she is in bed, and several hours have elapsed as if in a moment. There are countless variations on the theme, and while some accounts must be the products of fantasists and self-publicists' narcissism, it would be unfair to dismiss the genuine fear and long-term distress experienced by countless other "abductees". So, what's going on?

Health professionals often take the symptoms seriously, if not the reported causes. In 1992, the *Harvard University Gazette* published the findings of **Dr John Edward Mack** (1929–2004), a somewhat controversial psychiatrist at Harvard Medical School (and winner of the 1977

Pulitzer Prize for his biography of Lawrence of Arabia). Mack investigated hundreds of people claiming to be alien abductees. What struck him was their sanity and ordinariness. They comprised a classic cross-section of humanity; there were restaurant owners, secretaries, prison guards, college students, university administrators and so forth. There was nothing to suggest some particular pattern of craziness or unreliability among this diverse group of individuals. It's just that they all claimed, insistently, that they'd been meddled with by aliens.

Mack's excessive interest in aliens embarrassed Harvard Medical School. When he sought help from Harvard law professor and celebrity lawyer Alan Dershowitz, Harvard backed down from their most dire threats and reaffirmed his academic freedom to research whatever he liked, so long as he didn't "violate" their academic standards. Mack found that "as a group, abductees are no different from the general population in terms of psychopathology prevalence". He also noted fascinating parallels between twentieth-century alien abduction experiences and similar accounts, throughout history, of visions and eerie transportations. While the hardware and costumes shift through the ages, the basic story remains the same. A well-known artwork typifies the scene.

Strange encounters with mysterious interlopers have long been the stuff of lore and legend, as illustrated in Henry Fuseli's disturbing painting *The Nightmare*.

Painter Henry Fuseli (1741–1825) was fascinated by dreams and visions. His most famous work, *The Nightmare,* was unveiled at the Royal Academy, London, in 1782, and subsequently became popular as a commercial engraving. A goblin sits on the stomach of a sleeping woman, while a mad-eyed horse pokes its head around a curtain to survey the scene. The woman is wearing a thin nightdress, and her pose suggests sexual vulnerability. She swoons rather than slumbers. According to Tom Lubbock, art critic for *The Independent,* the most disturbing thing about the creature is that it isn't actually doing anything to the woman:

> "It's just sitting on her, inert. It has some calm and horrible purpose, which is worse. And it turns its bulging eyes to meet the viewer's in a way that shows a mind at work."

Fuseli drew his inspiration from John Milton's *Paradise Lost,* and various medieval legends. *The Nightmare* shows an **incubus** (from the Latin *incubare,* meaning "to lay upon"). This is a demon in male form who, according to many legends, lies upon sleepers, especially women, in order to have its wicked way with them. The female equivalent is a succubus. Interestingly, in most visual representations, incubi usually look like demons, while succubi tend toward the voluptuous *femme fatale* type. According to *Paradise Lost:*

> For spirits, when they please
> Can either sex assume, or both; so soft
> And uncompounded is their essence pure,
> Not tied or manacled with joint or limb,
> Nor founded on the brittle strength of bones,
> Like cumbrous flesh; but, in what shape they choose,
> Dilated or condensed, bright or obscure,
> Can execute their airy purposes,
> And works of love or enmity fulfil.

Most cultures apparently have equivalents. The Hmong, an Asian ethnic group from the mountainous regions of Vietnam, Laos and Thailand, talk of *dab tsog,* a child-sized demon sitting on their chests. In China it's *guǐ yā chuáng,* "ghost pressing on bed", while Japan adds an intriguing dash of hardware with *kanashibari,* meaning "trapped in metal". Everywhere we look, both through time and across geographical borders, eerie entities of short-to-medium height are paralysing helpless victims.

Sleep paralysis

One common link may be sleep paralysis. During normal rapid eye movement (REM) sleep, when most dreams happen, our muscles are deactivated so we don't physically act out our dreams. It's probable that the muscles themselves also need a period of inactivity. Even automatic reflexes, like kicking out when the knee is tapped, don't function during REM sleep. Occasionally when we wake up, there's a slight lag between the brain regaining consciousness and our ability to get those muscles moving again. We become aware of the fact that we can't move. This is known as **sleep paralysis**, and in most people it passes in a few moments. Sometimes, however, it can last for several minutes, creating intense psychological disorientation, and even triggering hallucinations when dream states persist after we wake. It's not surprising that some people may interpret this kind of physical and psychological experience as an alien abduction.

British psychologist Susan Blackmore began her career fascinated by extrasensory perception and the paranormal. Over the years, as evidence for these phenomena failed to stack up, her attitude moved from belief to scepticism. Today she is a fellow of the Committee of Skeptical Inquiry (CSI), an international organization promoting "scientific inquiry, critical investigation, and the use of reason in examining controversial and extraordinary claims".

In a November 1994 article for *New Scientist*, Blackmore explored the work of Michael Persinger, a neuroscientist at Laurentian University in Ontario, who used a helmet of electromagnets to disrupt the firing of brain cells in the temporal lobe. His volunteer subjects reported experiencing a range of emotions, sometimes ecstatic and spiritual, at other times profoundly disturbing. Blackmore gave the helmet a try, and said she felt,

> "As though two hands had grabbed my shoulders and were bodily yanking me upright ... If someone told me an alien was responsible and invited me to join an abductees' support group, I might well prefer to believe the idea, rather than accept I was going mad."

It's the specifically frightening nature of alien abduction scenarios that interests Blackmore. A particular twist is added by the fight-or-flight response (or acute stress response), in which the sympathetic nervous system immediately reacts to a perceived threat, priming an animal to make a snap decision about whether to fight a dangerous rival or run for its life. Sleep paralysis often creates the sensation that a sinister "other" presence is in the room, from which there is no escape. The fight-or-flight response is evoked at a moment when the muscles are still disabled from sleep, despite their owner being awake.

Blackmore is not alone in thinking that many abduction narratives owe their origins to biological and neurological causes, rather than extraterrestrial interventions. According to Al Cheyne, an associate professor of psychology at the University of Waterloo in Ontario, "Trolls or witches no longer constitute plausible interpretations of these hallucinations. The notion of aliens from outer space is more contemporary and plausible to the modern mind."

Alien-looking aliens

The strangest element uniting almost all "close encounter of the third kind" witness reports is that the aliens appear more or less human, or at least humanoid. The diminutive **Greys** have entered popular imagination,

The aliens of *Close Encounters of the Third Kind* (1977) are archetypal diminutive "Greys" – and they seem to have a fondness for twirling.

with their long arms, skinny bodies and large, dark, almond-shaped eyes perched in bulbous heads. In the early years of science fiction, many authors depicted aliens with oversized heads accommodating super-intelligent brains. Typically, these enlarged noggins were perched atop scrawny bodies, with stick-like arms and legs deemed barely necessary by creatures who prefer mind control over brawn. However, our modern notions of the body beautiful suggest that we seek physical improvements as well as mental ones, so we have no reason to think that aliens might surrender their fine physiques.

Even if aliens did happen to come from a world just like ours, they still wouldn't look like us, because of countless tiny variations in their evolutionary history. Suppose our ape ancestors hadn't been so good at survival, and modern humans had never evolved, or that the sudden K-T extinction event (see p.79) never happened, and the dinosaurs had inherited the Earth? Or who knows – perhaps meerkats could have ended up as the first terrestrial creatures to invent the wheel? If we think for a moment

Convergent evolution

Physical circumstances are similar for all plants and animals occupying particular niches of Earth's environment. In response, animals from different branches of the evolutionary tree have evolved similar physical characteristics, such as legs or wings, to cope with very similar problems: how to overcome the same gravity; how to swim through the same waters or fly through the same atmosphere, and so on. This is known as convergent evolution. The wings of bats, birds and insects evolved independently from each other, as did the eyes of mammals, squid, octopi and jellyfish, because propulsion through air is expedient for hunting and escape, and sensing electromagnetic radiation (light) is useful in a sunlit world. The terrestrial coconut crab may not share your fondness for the whiff of bacon, but it has evolved a sense of smell never-theless, because air and water alike convey chemical clues about the external environment, and it pays to pick up on those clues.

Aliens from a sufficiently Earth-like world may share at least some of our evolutionary traits, such as an ability to perceive distance via stereo vision, or a tendency to position sensitive eyeballs well away from organs of excretion. Symmetrical body shapes might also be prevalent throughout the galaxy, just as they are on Earth, for the simple reason that they are more physically stable than wonky ones. There are also aspects of molecular biology that create a bias towards symmetry. Aliens from a vaguely Earth-like world might exhibit recognizable symmetry, although "recognizable" is the key word here. A trained biologist would spot the symmetry in an alien shaped like a starfish. The rest of us wouldn't be so sure.

about how *alien* an alternative evolutionary history on this planet could have been, it becomes obvious that creatures from other worlds are not under any obligation to look like humans.

Aliens may not even be recognizable as distinct creatures. Many insects, including ants and termites, live in huge colonies containing thousands, or even millions, of individuals. A single ant responds in set ways to the behaviour of its fellow ants, and to chemical and visual cues. Put those ants together, however, and a complex and startlingly adaptive collective behaviour emerges. A single ant, with its tiny brain, cannot know very much about what is happening to the colony as a whole, but when all of them work in harness, each doing its own little bit, the colony acts as though it has an intelligent mind. It's amazing to see ants working together in huge teams to gather food or build nests, as though someone were controlling them. This is just one more illustration of how biology could deliver an intelligent alien species in countless non humanoid ways.

The bigger picture

It's impossible to look up at the night sky and not wonder what it all means. Where did the universe come from? Why does it exist? Is there some connection between Its external reality and the inner, conscious imaginations of its inhabitants?

The origins of the universe

According to accepted scientific theory, the universe is expanding because it still carries the momentum of a tremendous explosive event called the **Big Bang**, which happened in a single instant approximately 13.75 billion years ago. All the matter and energy in the universe blew outwards from a single dimensionless point. Elaborate mathematical descriptions of energy, gravity and matter interactions have been invented to calculate the sequence of events backwards in time to within fractions of a second of the Big Bang itself. Physicists display great confidence in their models.

Recent discoveries by space probes seem to support the Big Bang theory. All regions of the sky emit a faint microwave radiation. It comes from everywhere, yet from nowhere in particular. It's the warm afterglow of the initial explosion. The Cosmic Background Explorer satellite (COBE), and its successor, the Wilkinson Microwave Anisotropy Probe (WMAP), have found uneven patches in the radiation where tiny fluctuations in the initial fireball expanded to produce distortions in the fabric of space. Such distortions were crucial. Had the Big Bang been perfectly uniform, all the matter in the cosmos would have flown outwards without condensing into galaxies, stars and planets. Cosmologists have long believed that subatomic disturbances must have interfered with the too-tidy symmetry of the Big Bang within the first fractions of a second. Now COBE has found the microwave echo of those vital imperfections in creation. The

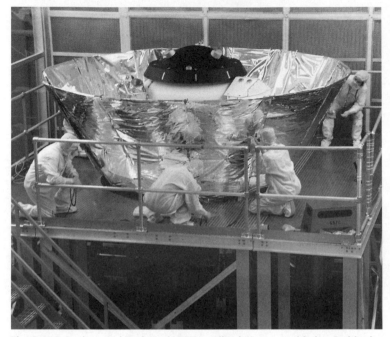

The Cosmic Background Explorer (COBE) satellite being assembled at Goddard Space Flight Center, before beginning its search for evidence of the Big Bang.

proponents of Big Bang are delighted. COBE's principal scientists, George Smoot and John Mather, were awarded the Nobel Prize for Physics in 2006 in honour of these important discoveries.

The more redshifted an object is, the older it must be, because it has travelled so much further outwards from the initial point of creation. But that mythical point is just a clumsy convenience of human language. Space and time have also expanded as an integral part of the universe; and since all points of the universe are expanding away from each other, there is no "centre" of the cosmos for us to discover. Mathematicians are happy to play around with these ideas, folding space into multiple dimensions and arguing that the shortest distance between two places in the cosmos isn't a straight line after all. They aren't so happy, though, with certain aspects of the Big Bang scenario.

A real whopper is that the gravitational attraction of all the stars and galaxies prevents the cosmos from breaking apart. Gravity is connected with mass, and it seems that the universe isn't massive enough to hold itself together. This isn't just a minor shortfall in the numbers: nine tenths of the necessary mass has yet to be accounted for. The current theory is that there may be colossal amounts of material in space that our instruments cannot detect, because it doesn't emit or reflect light or radio waves. This is the **Dark Matter hypothesis**. There are gaping holes in the Big Bang equations where the mass values from all this invisible material are supposed to fit.

Physicists such as Stephen Hawking also think the Big Bang model is flawed. He and many other prominent physicists champion a cosmos created from superstrings – quantum distortions in the fabric of the vacuum itself, which can spontaneously give rise to matter. Superstrings are presumed to be thinner than the width of a single atom, yet they give rise to immense energies. The universe they create does not start in a pinpoint fireball. It emerges out of what might usually be described as "nothing". Superstrings are clever mathematical inventions, designed to wish away the gravitational problems of Big Bang. They're probably the most advanced ideas the human mind has ever conceived. Yet the strings just create an even bigger question. They only work if the laws of quantum physics are somehow embedded in the vacuum. Those laws are very complicated. Certain versions of string theory require space-time to be entangled among 26 dimensions.

The point about superstrings and other such incredible theories is this: it doesn't matter that some of the theories disagree with each other, or that our knowledge is imperfect. That's only to be expected. All scientists agree that they are observing a universe that contains consistencies at every scale, from atoms to galaxies. The universe has order. It has *laws*. Human scientific theories are reflections of certain aspects of those laws; they are not a complete phrasing. So, where do those laws come from?

The fate of the universe

Astronomers are keen to work out what will happen to the universe in the distant future. Will it carry on expanding and eventually fizzle out like a depleted firework? Or will it gradually slow down, halt for a few billion years and then collapse inwards, creating an energetic Big Crunch that bounces outwards again? These questions disappear if the Big Bang concept is discounted altogether. British astronomer Fred Hoyle was a famous champion of the rival **Steady State theory**. This suggests that the expanding universe is constantly replenished from within by new material. Some modern cosmologists also suspect that there was no beginning, and there will be no end. There's no Crunch to come, and there never was a Big Bang.

The Anthropic Principle

For all its imperfections, our understanding of the cosmos, from stars and galaxies to the orbits of simple asteroids, has proved incredibly effective. Is this just a coincidence? This is where the **Anthropic Principle** kicks in. First proposed by Australian physicist Brandon Carter in 1973, and now widely discussed by almost all cosmologists, the principle has two variants: weak and strong. The Weak Anthropic Principle merely says that although the universe might seem suspiciously well-suited for life, we only notice this because we happen to exist within that universe. Had the universe turned out to be unsuitable, we would not be here to discuss the problem. End of problem. The weak version is favoured by many scientists and philosophers as the simplest way of eliminating pointless lines of inquiry. The Strong Anthropic Principle is much more interesting. It examines four fundamental forces of nature, and considers what might have happened if they had turned out differently.

The four fundamental forces of nature

Everything described in this book can be ascribed to this quartet of interactions:

▶ **Gravity**, which pulls matter towards matter.

▶ **Electromagnetism**, which provides atoms with their chemical characteristics at close ranges, especially their positively charged protons and negatively-charged electrons.

▶ **The strong interaction**, which binds the nucleus of an atom.

▶ **The weak interaction**, by which atoms decay radioactively under certain circumstances.

These four fundamental forces differ in strength, but the differences relate in the same way to all atoms in the cosmos. What would happen if any of these forces were to change, even slightly? Gravity is such a weak force at small scales that it scarcely counts. However, if the exact strength of gravity at the universe's dawn had been weaker by the smallest possible degree, the large-scale implications would have been dramatic. The galaxies would not have coalesced. Nascent suns wouldn't have reached ignition point. We wouldn't be here. On the other hand, if gravity had turned out a little stronger, then the rate of collisions between stars would have been so great that a typical solar system like ours wouldn't have survived long enough to produce stable planets and life. Again, we wouldn't be here.

What about electromagnetism? If the balance of the electromagnetic force altered in any way, all known chemistry would disintegrate. The same applies to the strong interaction. As for the weak interaction, it's tempting to think that we could do without radiation. However, if the weak interaction became slightly less weak, the stars could not burn and the elements of life could not be manufactured. These examples are gross simplifications. Any tiny change in the relationship between the four forces would produce not just the vexing inconveniences described above, but the dissolution of the entire cosmos as we know it. The four forces that make the universe possible can be found within every single atom. Describe the behaviour of an atom in terms of those forces, and the characteristics of stars, planets and galaxies follow with absolute inevitability. The same applies to biochemistry. Hawking's biographer Kitty Ferguson has observed, "If something in the cosmos isn't a product of the four forces, then it hasn't happened."

Now, getting back to the Strong Anthropic Principle. At the same instant our universe began, the laws of physics must also have been created, because the laws (the consistencies we observe) are an emergent property of the universe itself. There is no law-about-laws dictating that the four fundamental forces *had* to turn out the way they did. Physicists agree that a Big Bang could produce forces slightly different from the ones we actually find in the world. There are billions of possibilities of poor-quality balances between the forces. Any of these "bad" balances would have produced a universe incapable of maintaining order and creating life. So how did the Big Bang generate a "good" universe the first time round, against such colossal odds? Some interpretations of the Strong Anthropic Principle imply a purposeful (teleological) universe, because the chances of everything turning out badly from the Big Bang were so high that only a deliberate bias in the starting conditions could have altered the odds in its favour.

> **"We will never know completely who we are until we understand why the universe is constructed in such a way that it contains living things."**
>
> Lee Smolin, *The Life of the Cosmos* (1997)

These kinds of anthropic argument are tools for discussion, and nothing more. Most serious scientists would balk at any suggestions of deliberate design in the cosmos. Even so, astrophysicist Paul Davies is far from alone in suggesting that the raw data of the Big Bang, as we currently understand it, presents a profound puzzle, or what he calls a Goldilocks Enigma, that has yet to be resolved. His fellow physicists often joke about "the hand-set values", a casual shorthand to remind them that they still haven't explained *why* those four fundamental forces came into being so neatly packaged.

One possibility is that when a black hole reaches its point of infinite compression, it creates an equivalent "white hole" in another universe, spewing out matter in a Big Bang. Our universe emerged suddenly from a dimensionless point of infinitely compressed matter. A black hole is... a dimensionless point of infinitely compressed matter. To hijack a quote from Bart Simpson when confronted with a malfunctioning vacuum cleaner, "This does something I once thought was impossible. It sucks and it blows." Perhaps a black hole in an earlier universe gave birth to ours, while black holes in our universe are spawning yet more universes elsewhere?

Surely a black hole must swallow all the matter in one universe, before it could spit it out into another? Not necessarily. A black hole can compress a few million stars, or just the remnants of the one star it first came from, into a pinprick of infinitely condensed matter. Yet it takes only one such pinprick to create an entire universe.

The powerful gravity of a massive black hole (upper left) captures the gas being shed by its companion star in the Cassiopeia constellation (artist's concept).

If black holes are capable of creating new universes, this might explain how our particular universe could have arisen from a process quite similar to Darwinian evolution. In November 1996, two biologists, John Maynard Smith of the University of Sussex and Eörs Szathmary of the Collegium Budapest in Hungary, wrote an informal essay for the science journal *Nature* entitled "On the Likelihood of Habitable Worlds". Their central thesis has since been adapted by cosmologist Lee Smolin in his book *The Life of the Cosmos* (1997). This is how the theory works:

Our universe must be well-tuned for life because we are here to discuss it. Our universe also makes plenty of black holes, as a result of the same laws of physics that make life possible (the four fundamental forces). Those black holes, in turn, are spewing out plenty of other new universes. So, a universe that's good for black holes, and that's also suitable for life, is more likely to create new universes.

On the other hand, a universe that's *not* suitable for life is less likely to create black holes. So, it won't generate other universes. The more new universes that are created, the higher the chances that some of them will create their own black holes, and so on. The process is exponential, and life-suited universes will tend to predominate. And so, the dilemma of the Strong Anthropic Principle is reversed. Instead of being a one-off fluke, it's almost inevitable that we should find ourselves in a universe with a life-friendly set of physical laws and a stable balance between the four fundamental forces. It looks as though we may be faced with two opposed, but equally mind-boggling, possibilities:

▶ There is only one universe, and – against colossal odds – it is suspiciously well-suited for complex things such as stars, planets and living creatures to emerge.

▶ There are billions upon billions of universes in a "multiverse" spawned by black hole/white hole events. The multiverse has thrown the Big Bang dice billions of times.

Smolin has written that it might be possible, in the coming decade or two, to make specific astronomical observations in support of the **black hole–white hole theory**. In the meantime, the debate about multiple universes continues. This is no science-fiction backwater. It is a central question about the nature of reality that must be resolved. In April 2003, *The New York Times* ran an article by Paul Davies warning readers not to take multiple universes too seriously. He said that it isn't good science to speculate about things that are intrinsically unobservable, because there is no way to prove them. A month later, *Scientific American* published an article by physicist Max Tegmark asserting that parallel universes almost certainly *must* exist. Martin Gardner (1914–2010), renowned for his mathematical inventiveness, waded into the debate: "Surely the conjecture that there is just one universe and its Creator is infinitely simpler and easier to believe than that there are countless billions upon billions of worlds?"

The information universe

Physics seems to be showing us that everything we know of – from cats and dogs and trees to stars, planets and galaxies – is made from atoms and subatomic particles, in other words, from a relatively small set of fundamental building blocks. There may be a yet more fundamental unit, such as the superstring, that can account for all the universe's matter and energy. Even if we prove the existence of superstrings, we might be left with a more subtle question than "what is the universe made of?" How do we account for the *differences* we observe between things when they are all made from the same stuff?

One possible answer is that the cosmos may behave like a computer, using a simple code of subatomic components and forces, and combining them according to the rules of its software – what we call the laws of physics – to deliver many different outputs, just as a computer manipulates simple on–off pulses (zeros and ones) to create apparently complex 3D images in a virtual reality system. Scientists are now debating the potential importance of information, rather than matter and energy, as the true mechanism at the heart of existence. Any hydrogen atom, electron, proton or whatever is exactly the same (give or tak isotopic variations) as all of its cousins, no matter that there are countless tetratrillions of them in the universe as a whole. They're just pieces of information, bits of code. Theoretical physicist John Wheeler (1911–2008) said that we should "regard the physical world as made of information, with energy and matter as incidentals".

Dutch theoretical physicist Gerard 't Hooft, joint winner of the 1999 Nobel Prize for Physics, thinks that we live in a **holographic universe**. A 2D information structure shimmers on the cosmological horizon. The third spatial dimension that we observe is an illusory product of our tiny, human-scaled perspective. Apparently solid constructions, such as people and planets, may be nothing more than cosmic projections

Making three dimensions from two

An echo of Gerard 't Hooft's complicated idea is found in the kind of hologram we see in a gallery or science expo. It's made by firing a laser through an optical splitter, which divides the beam into two, creating two wide swathes of light rather than narrow beams. One swathe illuminates the subject, then the reflected light lands on a sensitive photographic plate. The other swathe is diverted straight to the plate, without glancing off the subject. When the plate is processed, the result is an interference pattern, rather than a picture.

The next trick is to shine the right kind of light at the plate, at which point a 3D image of the original subject can be seen. The strangest quality of a hologram is that each small fragment of the plate holds all the information required to reconstruct the entire 3D image. If you cut a hologram in half, then cut the halves in half again, and so on, you can still see the whole image in each piece. The "whole in every part" quality of a hologram may be a subtle clue about the universe's construction.

emerging from that far-off 2D horizon, as malleable and transitory as any computer's temporarily stored 3D models brought briefly to apparent life on a plasma screen. Imagine that screen not as a flat surface on your laptop, but as something that really does seem three dimensional, and you have a flavour of how the universe that we *think* we see and feel might be a virtual construct.

One key quality of a computer is that the information inside it has to be manipulated by some kind of physical architecture: a microprocessor chip, for instance, with its millions of transistor switches and memory modules for storing the results of all its calculations. There is some confusion among information theorists about whether or not information has any real meaning in the absence of such a system to process it, and just as importantly, something – or someone – to read it. If the cosmos is similarly information-based, what is the hidden architecture that stores all the information, manipulates it and outputs the results? And who's watching the output, interpreting it and judging it to have some kind of meaning?

From our point of view, observing matter and energy with our scientific instruments is roughly the equivalent of gazing through a powerful magnifying glass at the tiny pixels on the screen of a vast computer. We can learn a lot about the wonderful images that appear on the screen when all the pixels combine to make pictures. What we have yet to understand is what makes the pixels. We need to understand the processing architecture behind the screen of perceived reality, and how the cosmos creates its surface appearances. We can speculate that an advanced alien civilization might have access to that deeper level of cosmic architecture, just as our quantum physicists understand matter and energy at a deeper level than, say, Galileo.

These ideas might seem far-fetched, like something out of *The Matrix* movie, until we remember that we have already learned to simulate most kinds of familiar visual and sound experiences using nothing more than simple binary code computers. In all likelihood we are less than a couple of decades away from creating illusions indistinguishable from our current notions of reality. For the time being, we can step outside our arcade simulators at the end of a day's play and re-enter what we think of as the real world. There may be no such obvious exit from the greater cosmic illusion all around us.

If a tree falls in the forest...

Here's a puzzle. Lego bricks are famous for the way they can be stacked together into a limitless variety of shapes. You can build an aeroplane, a spaceship or a house, yet the bricks come in a very limited set of types. Let's imagine a Lego kit with 24 kinds of brick, just like the 24 fundamental subatomic units (electrons, photons, quarks etc) that are the building blocks of the universe. Now, suppose you build a beautiful life-sized cathedral using billions of tiny bricks to build up the finest details. You live a long life, then die and the model passes to your children, and their children. Then it languishes in an open-air museum for hundreds of years. Finally, somehow, it survives long after all human culture has vanished in a terrible calamity, and there's no one left to admire it, or to understand the history of its construction, nor what it was supposed to mean when it was built all those aeons ago. Is it still a beautiful model of a cathedral or is it a meaningless pile of billions of bricks in 24 different varieties?

There are many scientists who ask: does the universe exist independently of mind, or do creatures such as ourselves need to be aware of it in order for it to exist? If this sounds like a silly question, it's just a simple description of a problem that has vexed the quantum science community since the 1920s. The founders of quantum physics insisted that we actually make the universe real by interpreting its information content and giving it meaning. Niels Bohr, co-inventor of quantum mechanics, said, "A physicist is just an atom's way of looking at itself." In his 1992 book, *Gravitation*, John Wheeler expressed something similar:

> "May the universe in some sense be brought into being by the participation of those who participate? Participation is the new concept given to us by quantum mechanics. It strikes down the objective observer of classical theory, the person who stands safely behind the thick glass wall and watches what goes on without taking part."

According to Paul Davies, "Through conscious beings, the universe has generated self-awareness. This can be no trivial detail, no minor by-product of mindless, purposeless forces. We are truly meant to be here." In a similar vein, Brandon Carter declared: "Although our situation is not necessarily central, it is inevitably privileged to some extent." Certainly it is privileged if we are the only creatures in all the vastness of the cosmos who know how stars are formed, or what a galaxy is, or how subatomic particles are interlinked with supernovae. Davies uses the word "we" to encompass sentient entities anywhere, whether human or not.

In the words of Arthur C. Clarke again, "Sometimes I think we're alone in the universe, and sometimes I think we're not. In either case the idea is quite staggering." No wonder we're so keen to settle this question. Being solely responsible for knowing about the universe's existence is quite a burden.

On the other hand, perhaps we should stop relying so much on the heavens for our spiritual and emotional sense of direction, and get used to standing on our own feet. Maybe it's time to dispense with gods and aliens alike. In Jermoe Agel's *The Making of 2001* (1972), Stanley Kubrick said:

> "The most terrifying fact about the universe is not that it is hostile, but that it is indifferent; but if we can come to terms with this, and accept the challenges of life within the boundaries of death, then our existence can have genuine meaning. However vast the darkness, we must supply our own light."

The end and the beginning

By weight, our bodies are ten percent hydrogen, sixty percent oxygen and twenty percent carbon. The last ten percent is taken up principally by atoms of nitrogen, calcium, phosphorous, sulphur, sodium, magnesium, iron and copper. All these atoms were created not in our sun, but by other stars long since dead. They are not *our* atoms. They do not belong to us, nor to our bodies. They are just passing through us for a very short while. They have existed for billions of years already, and will continue to exist

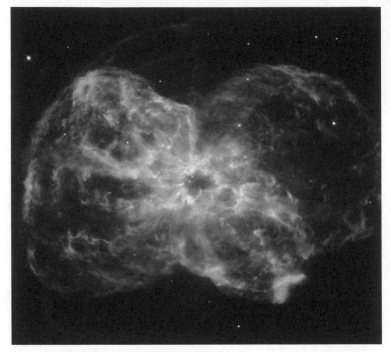

Death of a sun-like star: the Hubble Space Telescope took this picture of a dying star in its white giant phase (centre), casting off its outer layers of gas.

long after we are gone. They have barely begun their chemical adventures, and will combine and recombine long after Earth itself has ended its days as a life-giving world.

Our sun has burned for five billion years, and will burn for five billion more; after which, it will have used up a crucial portion of its available hydrogen fuel. Eventually, the sun will become an unpredictable monster, prone to sudden nuclear shutdowns and reignitions, expansions and contractions, alternating over several millennia. It will end its days as a **red giant** – so gigantic, in fact, that its wayward mantle of gases will swallow up half the planets in the solar system, including Earth. Everything on our world that lives and breathes will be utterly vaporized in a slow but remorseless firestorm. The grass, the trees, even the toughened lichen that clings to granite will be blasted into atoms. The oceans will boil away to the last drop. The ice caps will vaporize. The sands on every beach in the world will fuse into glass. Mercury and Venus will probably vanish altogether.

If Earth survives the storm, it will remain only as a scorched lump of rock. The sun's final life-giving service will be to vaporize the thin sliver of biosphere on the Earth's surface, blasting it into space once more so that its atoms might one day contribute towards new worlds and new forms of life. Perhaps an intelligent and curious creature, born many billions of years from now on an alien world containing a few of our second-hand atoms, will wonder where those atoms came from.

None of these grand-scale events can conceivably matter to us, our descendants or to the cyborg creatures that we may become after our purely biological story has ended. What will count are the decisions that we make during this current, extraordinarily brief spark of time during which the benefits of our intelligence are balanced almost exactly by the environmental stresses caused by that same intelligence. To put it simply, we may be too clever for our own good, yet not nearly clever enough. The balance will tip, one way or another, in decades or centuries, not aeons. If we think like sentient occupants of an infinite cosmos instead of nervous beasts scrabbling from day to day on the surface of a tiny, overcrowded ball of rock – if we aspire to live at least as long as our own star continues to shine – we might have a chance of becoming creatures worth meeting.

Resources

It seems there are as many books, websites and movies about aliens as there are stars in the universe. For those of us who want a solid scientific ground to stand on while we look up at the night sky in wonderment – and, equally, relish a good old fashioned sci-fi spectacular – here is a selection of material.

Books

Captured by Aliens: The Search for Life and Truth in a Very Large Universe Joel Achenbach (Citadel Press, 2003)
This volume makes uncomfortable reading for those of us with a romantic or spiritual streak. Unfortunately, things that we wish to be true may not be. It seems we humans are very good at fooling ourselves. After reading this, you'll have your wits about you when it comes to assessing UFO stories.

Seven Clues to the Origin of Life: A Scientific Detective Story A.G. Cairns-Smith (Cambridge University Press, 1990)
We take mud for granted, but perhaps we shouldn't. Its clay minerals may have given life its first essential helping hand on the early Earth.

The Eerie Silence: Are We Alone in the Universe? Paul Davies (Allen Lane, 2010)
With the exception of the quite deliriously brilliant book you're holding in your hand right now, this is the one you have to have. Davies is chair of the SETI post-detection task group, so he knows a thing or two. His lucid, elegant prose is thought-provoking and balanced. His explorations will satisfy alien enthusiasts and "we are alone" hard-liners alike.

Life on Other Worlds: The 20th-Century Extraterrestrial Life Debate Steven J. Dick (Cambridge University Press, 2001)
Dick is a senior NASA historian, and this book demonstrates why. His academic rigour is unbending, yet he never lapses into the humourless, pedantic prose of a university lecturer. This is a definitive source for

those interested in the deeper cultural background to SETI, from the time of the ancient Greeks to the present day.

Is Anyone Out There? The Scientific Search for Extraterrestrial Intelligence Frank Drake & Dava Sobel (Delta, 1994)

The world's foremost SETI scientist teams up with an expert popular science writer for a journey of cosmic and personal exploration. It's a few years old now, but indispensible reading nevertheless.

Unmasking Europa: The Search for Life on Jupiter's Ocean Moon Richard Greenberg (Springer, 2008)

A lively and fascinating account of an icy little moon that could one day turn out to be much more fascinating than Mars. We already believe that its frozen crust hides a deep ocean. Greenberg suggests that the surface ice may be even thinner than NASA thinks.

Other Worlds: the Search for Life in the Universe Michael D. Lemonick (Simon & Schuster, 1998)

This book is actually more about the craft of planet-hunting than the quest for extraterrestrial life but it's very gripping, with colourful accounts of the people and personalities involved in this new branch of astronomy.

Cosmos Carl Sagan (Abacus, 1983)

This is a benchmark in popular science writing. Don't worry if some of the details in this thirty-year-old book are out of date. All the important stuff remains current, from the origins of the Earth to the role of intelligence in the galaxy. Those wanting a narrative thrill should read Sagan's *Contact* (1985), in which he wraps his favourite SETI ideas into a very believable novel.

Confessions of an Alien Hunter: A Scientist's Search for Extraterrestrial Intelligence Seth Shostak (National Geographic Society, 2009)

Despite having spent much of his career fruitlessly hunting extraterrestrial intelligences, Shostak is as eager for the chase as the day he began. This is a fascinating and accessible look behind the SETI scenes.

The Life of the Cosmos Lee Smolin (Phoenix, 1997)

Does Darwinian selection operate at the grandest sacks of existence? Are black holes breeding conduits for a multiplicity of universes? It may sound insane, but all those stars and galaxies had to come from somewhere. Smolin makes a compelling case that our universe may not be the only one.

The Scientific Exploration of Mars
Frederic Taylor (Cambridge University Press, 2010)
Written by a scientist who actually studies the Red Planet, this is a comprehensive account of Martian investigations from the nineteenth century to the present day.

Landscapes of Mars: A Visual Tour
Gregory L. Vogt (Springer, 2008) No matter how many books and articles we read about Mars, the planet only becomes compellingly real when we see pictures of its surface, and here it looks as crisp and clear as if it were just a few streets away.

Biological Big Bang: Panspermia & Origins of Life
Chandra Wickramasinghe (Cosmology Science Publishers, 2011)
Edited by the long-time academic collaborator of controversialist Fred Hoyle, this book is a collection of essays from various scientists, all supporting panspermia – the idea that life on Earth was seeded from space.

The Case for Mars: The Plan to Settle the Red Planet and Why We Must Robert Zubrin (Free Press, 2011)
Revised and updated for a new edition, this no-holds-barred polemic against "business as usual" NASA clunkiness presents a simpler and cheaper way of reaching the Red Planet. It's so convincing, it makes you think we could send a mission there next year.

Technical papers

Online versions of the historic SETI-related scientific papers mentioned in this book are well worth a look, especially for technically minded readers.

The Discovery of ETI as a High-Consequence, Low-Probability Event Iván Almár & Jill Tarter (Acta Astronautica, Vol. 68)
This is the paper that led to the Rio Scale, a statistical tool for assessing the cultural and political consequences of an alien contact, based on similar models for asteroid impacts.
www.setileague.org/iaaseti/abst2000/almar.pdf

The Post-Detection SETI Protocol John Billingham (International Academy of Astronautics Position Paper, 22 March 1996)

What if we were to pick up an alien signal? Who among us could be trusted with the knowledge, and when should the rest of the world be told? www.coseti.org/setiprot

See also the current draft of the protocol itself: "Declaration of Principles Concerning the Conduct of the Search for Extraterrestrial Intelligence" www.setileague.org/iaaseti/protocols_rev2010.pdf

Alien Abductions, Sleep Paralysis and the Temporal Lobe Susan Blackmore & Marcus Cox (European Journal of UFO and Abduction Studies, 2000, pp.113–118)

Can alien encounters be explained by neurological phenomena inside our own heads? It's a more likely explanation than most.

Searching for Interstellar Communications Giuseppe Cocconi & Philip Morrison (Nature, Vol. 184, No. 4690, pp.844–846, 19 September 1959)

This is the seed from which the oak tree of modern radio SETI was born. www.coseti.org/morris_0.

The Big Ear Wow! Signal Jerry R. Ehman (30th Anniversary Report)

A very detailed analysis of the 1977 data that electrified a young radio astronomer, and which has puzzled him ever since. Arguments against accidental terrestrial sources for the signal are presented in strict scientific terms. www.bigear.org/Wow30th/wow30th

Movies

Science-fiction and fantasy fans have a wealth of alien-themed movies to choose from. Most are fun, and some are good. John Scalzi's *Rough Guide to Sci-Fi Movies* (Rough Guides, 2005) is an essential primer in the classics of the genre, including all the alien-related movies that are more about shocks and thrills than genuine scientific speculation.

Three movies in particular stand out from the crowd as thoughtful commentaries on SETI and the possibilities of alien contact. In this author's ranking of preference and cultural importance, they are:

2001: A Space Odyssey (Stanley Kubrick, 1968)

This is the benchmark space movie. Its hardware still looks stunningly convincing, nearly half a century after the designers first sat down at their drawing boards. In an age before colour TV, Kubrick's boffins conjured up the illusion of flatscreen computer read-outs and other twenty-first-century toys. Mysterious aliens, and the prospects of machine intelligence, are woven into a satisfyingly obscure vision of our place in the cosmos.

Close Encounters of the Third Kind (Steven Spielberg, 1977)

The set-up feels right. Yes, the big alien mother ships arrive, but instead of invading, their aim is to make contact. Government goons and scientists do what we'd expect: set up a secret reception committee, expecting to tell the world later about this momentous event. It's just that the aliens haven't restricted their "hello" messages to official folk, so the welcoming party is gatecrashed by wide-eyed civilians.

Contact (Robert Zemeckis, 1997)

Based on the novel of the same name by Carl Sagan, this is an intelligent and wide ranging exploration of radio-based SETI themes. The science is entirely credible, at least until the climax, where we get a spectacular pay-off and what seems like an alien encounter. But the last few minutes of the movie send us back to square one in an intriguing way.

Websites

Big Ear Memorial Website www.bigear.org

Sadly, the radio telescope that picked up the most compelling candidate for an alien transmission – the 1977 "Wow!" signal – no longer exists. This website keeps its memory, and research data, alive.

European Space Agency www.esa.int

Keep tabs on the ESA's science telescopes and interplanetary probes.

Japan Aerospace Exploration Agency www.jaxa.jp

The Japan Aerospace Exploration Agency (JAXA) has pulled off some impressive asteroid rendezvous missions in recent years.

Jet Propulsion Laboratory www.jpl.nasa.gov

Almost all NASA's orbiting space telescopes and robotic deep-space and planetary landing missions are conducted by JPL in Pasadena, California. It's a small yet exceptionally productive operation managed by the nearby California Institute of Technology (Caltech) on behalf of

NASA. JPL has sent probes to all the solar system's planets and most of the major moons. It operates a fleet of Mars-orbiting surveyors and wheeled surface rovers. Check here first for news about specific missions.

Jodrell Bank Centre for Astrophysics www.jb.man.ac.uk

Britain has one of the world's largest radio astronomy dishes, operated by Manchester University from a leafy field near Macclesfield in Cheshire. Threatened with funding cuts a few years ago, the facility has at last been rescued and revived.

NASA Astrobiology Institute www.astrobiology.nasa.gov/nai

The NAI was established in 1998 as a virtual, distributed alliance of research efforts in collaboration with other national and international science communities. This is where the bug-hunters and pre-biotic-broth mixers congregate. The "teams" button on the NAI homepage links you to fourteen relevant partner organizations, all conducting astrobiology research.

National Radio Astronomy Observatory www.nrao.edu

From the Drake equation and Project Ozma and beyond, this is the Big Daddy of radio astronomy outfits. SETI is just one facet in a major tranche of NRAO astronomy programmes.

National Security Agency, UFO Documents Index
www.nsa.gov/public_info/declass/ufo/index

If all the UFO documentation in America was stacked up, it would probably reach to the moon. The US government's spookiest intelligence agency has released plenty of material in response to numerous requests.

Seti At Home www.setiathome.berkeley.edu

Be the first to identify a signal from an ETI! Your chances of such fame are slender. Even so, it's less likely that anyone's going to find anything unless a lot of people keep looking – so join in.

SETI Institute www.seti.org

The one and only true home of the alien hunters, where our fondest philosophical hopes and dreams of contact are matched by disciplined science and intellectual rigour. Senior astronomer Seth Shostak's many blogs and essays are always fascinating, but there's plenty of SETI history here, too, going back to the Ozma days.

Seti Quest www.setiquest.org

Everything's got an app these days. The SETI Institute wants to harness the creative power of the masses in a bid to create open-source software that might improve upon current SETI search strategies. You may not be the first to detect an alien signal, but perhaps you could be one of those who writes the code that finds it.

Space www.space.com

This indispensible site keeps tabs on the latest space missions, future plans and current scientific controversies in planetary science and cosmology. A dedicated button at the top of the homepage, "Search for Life", focuses specifically on astrobiology-related news and SETI updates.

UK National Archives: Unidentified Flying Objects
ufos.nationalarchives.gov.uk

Whatever your feelings about flying saucers, the newly released UK government archives do contain a few intriguing mysteries, and some embarrassing admissions – such as the fact that shortage of funds prevented officials from investigating some of the more interesting reports.

University of California, Berkeley seti.berkeley.edu/resources

UC Berkeley hosts a great website leading you to other useful SETI-based links.

US National Archives: Unidentified Flying Objects
www.archives.gov/foia/ufos

The motherlode for UFO enthusiasts, rich in myth and secrecy, with enough material to keep us in movie plots for generations to come.

Woods Hole Oceanographic Institution www.whoi.edu

Marine biologists dream of being posted here – Massachusetts's glamorous Cape Cod peninsula. The WHOI is a non-profit ocean research organization with its own fleet of ships and deep-diving submarines, including the Alvin, famous for researching deep sea "smokers".

Picture credits

Chapter dividers

Index

U

T

V